Frontiers in Clinical Drug Research - Anti-Cancer Agents
(Volume 5)

Edited by

Atta-ur-Rahman, *FRS*
Honorary Life Fellow,
Kings College, University of Cambridge, Cambridge, UK

Frontiers in Clinical Drug Research - Anti-Cancer Agents

Volume # 5

Editor: Atta-ur-Rahman

ISSN (Online): 2215-0803

ISSN (Print): 2451-8905

ISBN (Online): 978-981-14-0515-0

ISBN (Print): 978-981-14-0514-3

©2019, Bentham eBooks imprint.

Published by Bentham Science Publishers Pte. Ltd. Singapore. All Rights Reserved.

need for a court order if at any point you breach any terms of this License Agreement. In no event will any delay or failure by Bentham Science Publishers in enforcing your compliance with this License Agreement constitute a waiver of any of its rights.

3. You acknowledge that you have read this License Agreement, and agree to be bound by its terms and conditions. To the extent that any other terms and conditions presented on any website of Bentham Science Publishers conflict with, or are inconsistent with, the terms and conditions set out in this License Agreement, you acknowledge that the terms and conditions set out in this License Agreement shall prevail.

Bentham Science Publishers Pte. Ltd.
80 Robinson Road #02-00
Singapore 068898
Singapore
Email: subscriptions@benthamscience.net

CONTENTS

PREFACE

Frontiers in Clinical Drug Research - Anti-Cancer Agents presents recent developments in various therapeutic approaches against different types of cancer. The book is a valuable resource for pharmaceutical scientists, postgraduate students and researchers seeking updated and critical information for developing clinical trials and devising research plans in anti-cancer research. The chapters are written by authorities in the field. The contents of this volume represent exciting recent researches on Acute Myeloid Leukemia, Chemotherapy, Gastrointestinal Cancer, Anti-cancer Therapy, Breast Cancer Cells, and Lung Cancer. I hope that the readers will find these reviews valuable and thought provoking so that they may trigger further research in the quest for the new and novel therapies against cancers.

I am grateful for the timely efforts made by the editorial personnel, especially Mr. Mahmood Alam (Director Publications), and Mr. Shehzad Naqvi (Editorial Manager Publications) at Bentham Science Publishers.

Atta-ur-Rahman, *FRS*
Honorary Life Fellow
Kings College
University of Cambridge
UK

List of Contributors

Atsuhiro Tsubaki	Institute for Human Movement and Medical Sciences, Niigata University of Health and Welfare, Niigata, Japan
Andrew J. Sanders	Cardiff China Medical Research Collaborative, Cardiff University School of Medicine, Cardiff University, Heath Park, Cardiff, UK
Chang Gong	Guangdong Provincial Key Laboratory of Malignant Tumor Epigenetic and Gene Regulation, Breast Tumor Center, Sun Yat-sen Memorial Hospital, Sun Yat-sen University, Varna, China
David Kerr	Radcliffe Department of Medicine, University of Oxford, UK
Etienne Paubelle	Hematology Department, University of Arizonaospices Civils de Lyon, Lyon-Sud Hospital, Pierre Bénite, France
Işıl Yıldırım	Beykent University, Vocational School, The University of Hong Kong, Istanbul, Turkey
Jack B. Fu	Department of Palliative, Rehabilitation Integrative Medicine, University of Texas MD Anderson Cancer Center, Houston, TX, USA
Kenneth K.W. To	School of Pharmacy, Faculty of Medicine, The Chinese University of Hong Kong, Hong Kong SAR, China
Onyinyechi Duru	Department of Oncology, Nottingham University Hospital, City Campus, Nottingham, UK
Shinichiro Morishita	Institute for Human Movement and Medical Sciences, Niigata University of Health and Welfare, Niigata, Japan
Wen G. Jiang	Cardiff China Medical Research Collaborative, Cardiff University School of Medicine, Cardiff University, Heath Park, Cardiff, UK
Wing-Sum Tong	School of Pharmacy, Faculty of Medicine, The Chinese University of Hong Kong, Hong Kong SAR, China
Xavier Thomas	Hematology Department, University of Arizonaospices Civils de Lyon, Lyon-Sud Hospital, Pierre Bénite, France
Yuequan Shi	China Medical University, Liaoning, China
Zifang Zou	China Medical University, Liaoning, China
Zihao Liu	Guangdong Provincial Key Laboratory of Malignant Tumor Epigenetic and Gene Regulation, Breast Tumor Center, Sun Yat-sen Memorial Hospital, Sun Yat-sen University, Varna, China

Second-Generation Protein Kinase Inhibitor – A Focus on Quizartinib, A Promising Targeted Therapy for High-Risk FLT3[+] Patients with Acute Myeloid Leukemia

Xavier Thomas[*] and **Etienne Paubelle**

Hematology Department, Hospices Civils de Lyon, Lyon-Sud Hospital, Pierre Bénite, France

Abstract: Fms-like tyrosine kinase 3 (FLT3) is one of the most commonly mutated genes in acute myeloid leukemia (AML). While first-generation FLT3 tyrosine kinase inhibitors are relatively non-specific for FLT3 with other potential targets, the next-generation inhibitors appear more potent and selective. Among them quizartinib is the most clinically advanced. The greater potency and selectivity of this drug promises greater efficacy and less toxicity in FLT3-mutated AML. It is currently studied across virtually all disease settings, and its use in combination with chemotherapy appears promising in FLT3[+] patients. In this review, we summarize the current data on quizartinib and the encouraging clinical data that have also emerged with other second- or further-generation FLT3 inhibitors, after recalling results observed with first-generation inhibitors.

Keywords: Acute Myeloid Leukemia, Chemotherapy, c-Kit Inhibition, FLT3 inhibitors, Prognosis, Quizartinib, Relapse, Targeted Therapy, Treatment, Tyrosine Kinase Inhibitors.

INTRODUCTION

Acute myeloid leukemia (AML) is characterized by aberrant proliferation of myeloid progenitor cells that have lost the ability to differentiate into mature cells. It represents a group of hematopoietic stem cell malignancies, affecting approximately 2 to 3 adults per 100 000 each year in Western countries. Outcomes for the majority of patients remain poor. Cytogenetic aberrations represent one of the most important independent prognostic factors in AML. Over the past decade, significant progress has been made in the understanding of the

[*] **Corresponding author Xavier Thomas:** Hematology Department, Hospices Civils de Lyon, Lyon-Sud Hospital, Bât. 1G, 165 chemin du Grand Revoyet, 69495 Pierre Bénite, France; Tel: +33478862235; Fax: +33472678880; E-mail: xavier.thomas@chu-lyon.fr

Atta-ur-Rahman (Ed.)

cytogenetic and molecular determinants of AML pathogenesis. Novel treatment strategies take into account these prognostic factors to develop risk-stratified treatment options in order to offer more adapted treatment options to patients at high risk.

The National Comprehensive Cancer Network (NCCN) guidelines for AML induction chemotherapy include a 3-day administration of an anthracycline (idarubicin 12 mg/m^2 or daunorubicin 45 to 90 mg/m^2 as a short IV infusion) combined with a 7-day administration of standard dose cytarabine (100 to 200 mg/m^2 as continuous IV infusion) ("3 + 7" chemotherapy regimen) [1]. With this schedule, morphological complete remission (CR) rates are about 70 to 80%, depending on age and patient selection [2 - 5]. Recommended post-remission chemotherapy for AML includes consolidation with high-dose cytarabine [1]. Cytarabine is given at 2 to 3 g/m^{2}in younger adults and 1 to 1.5 g/m^2 in patients ≥ 60 years, administered as an IV infusion every 12 hours on days 1, 3, and 5 for a total of 6 doses.

The growth and differentiation of hematopoietic cells is governed by the concerted action of growth factors and their receptors. FMS-like tyrosine kinase 3 (FLT3) is a transmembrane tyrosine kinase that belongs to the class 3 split-kinase domain family of receptor tyrosine kinase [6]. The gene is located at chromosome 13q12. FLT3 is expressed in AML cells in about 90% of cases and stimulates survival and proliferation of leukemic blasts [7]. FLT3 is mutated in about 30% of AML cases [8, 9]. Two major classes of activating mutations have been identified: internal tandem duplications (ITDs) (24%) which affect the juxtamembrane domain of the receptor [10] and point mutations in the activation loop, with a majority at the D835 residue (7%) [8, 11]. Internal tandem duplication of the FLT3 gene leads to constitutively activated receptor tyrosine kinase and its downstream signaling pathways, which in turn leads to dysregulation of cellular proliferation and enhanced cell survival [12]. This activation of signaling pathways is important in the pathogenesis of AML. Patients with FLT3-ITD mutations have a worse prognosis than those with wild-type FLT3, due to a higher relapse rate [13, 14]. The presence of FLT3-ITD mutations is widely accepted as a poor prognostic factor in cytogenetically normal AML. A high allelic ratio of FLT3-ITD/wild type was associated with inferior survival [15]. However, another study showed a benefit for patients with a high allelic ratio after allogeneic hematopoietic stem cell transplantation (HSCT) in first CR [16]. Intriguingly, FLT3-tyrosine kinase domain (TKD) mutations have generally not been associated with similar degree of negative prognosis [13]. Because of technical limitations and instability of FLT3 as marker, the clinical application for minimal residual disease (MRD) assessment is currently limited [17].

Currently, no FLT3 inhibitors have been approved for AML. There is currently no consensus recommendation for a consolidation cytarabine regimen for FLT3$^+$ AML patients. However, the European LeukemiaNet suggests to use allogeneic HSCT as a consolidation for patients with FLT3-ITD who are 18 to 60 years of age and intermediate-dose cytarabine for consolidation in patients over the age of 60 [18]. Inhibition of the receptor tyrosine kinases using small molecules represents an attractive therapeutic target. In the last decade, several molecules with activity against FLT3 have been tested. The relative nonselectivity of some of these agents, and suboptimal pharmacokinetics associated with others has led to unimpressive results. More recently, more selective and more potent FLT3 inhibitors have been developed (Table **1**). Among them, quizartinib represents the more clinically advanced next-generation FLT3 inhibitor in AML. It corresponds to a class III receptor tyrosine kinase inhibitor exhibiting highly potent and selective inhibition of FLT3. Initial studies with quizartinib were very promising based on a high response rate when administered as monotherapy and promising combination studies with chemotherapy are ongoing.

Table 1. FLT3 inhibitors and their targets.

FLT3 inhibitor	Targets
First-generation FLT3 inhibitors	
Sunitinib	FLT3, KIT, KDR, PDGFR
Lestaurtinib	JAK2, FLT3, TrK A
Tandutinib	FLT3, PDGFR, c-KIT
Sorafenib	FLT3, c-KIT, VEGFR, PDGFR, RAF-1
Midostaurin	FLT3, c-KIT, PDGFRb, VEGFR
Next-generation FLT3 inhibitors	
Quizartinib	FLT3, c-KIT, PDGFRa
Crenolanib	FLT3, PDGFR
Gilteritinib	FLT3, AXL
Pacritinib	JAK2, FLT3, IRAK1, cFMS/CSF1R
Ponatinib	BCR/ABL, FLT3, c-KIT, FGFR1, PDGFRa

Abbreviations: ABL, Abelson; BCR, Breakpoint cluster region; CSF1R, colony-stimulating factor-1; FGFR1, Fibroblast growth factor receptor 1; FLT3, Fms-like tyrosine kinase 3; IRAK1, interleukin-1 receptor-associated kinase; JAK2, Janus kinase 2; PDGFR, platelet-derived growth factor receptor; TrK A, Tropomyosin receptor kinase A; VEGFR, vascular endothelial growth factor receptor.

This review mainly focuses on quizartinib that currently appears as the most potent and specific FLT3 inhibitor, but the review also reports on initial results obtained with first-generation tyrosine kinase inhibitors and on the more recent promising ones described with the new further-generation inhibitors.

FIRST-GENERATION FLT3 TYROSINE KINASE INHIBITORS

These first-generation FLT3 inhibitors (midostaurin, lestaurtinib, sunitinib, sorafenib, and tandutinib) are multi-kinase inhibitors. They showed unfavorable pharmacokinetics with high plasma protein binding, unpredictable free drug levels, induced metabolism, and prominent gastrointestinal toxicity, and significant drug-drug interactions, especially with azole antifungals. Some, such as sunitinib and lestaurtinib, have not demonstrated sufficient promise for further development. Most are not inhibitors of FLT3 specifically, but often have multiple kinase targets. With the exception of sorafenib, they also demonstrated only limited activity as single agents with best responses on peripheral blast clearances, but showed interesting results in combination with intensive chemotherapy.

Sunitinib (SU11248) has been approved for renal cell carcinoma, gastrointestinal stromal tumor, and neuroendocrine tumor. A phase 1 study of sunitinib single agent therapy in patients with refractory or relapsed AML showed short CR durations as well as significant toxicities [19]. A subsequent phase 1/2 study of sunitinib combined with intensive chemotherapy showed CR rates of 59% in older patients harboring activating FLT3 mutations with a median survival of 18.8 months and a median relapse-free survival of 11 months [20].

Lestaurtinib (CEP-701) is an indolocarbazole derivative found to have potent *in vitro* activity against FLT3 [21]. Early phase studies in humans showed that lestaurtinib was highly bound to plasma protein, specifically α1-acid glycoprotein. Lestaurtinib was well tolerated and induced reductions in peripheral blood and marrow blasts in patients with FLT3 activating mutations [22 - 24]. *In vitro* studies of lestaurtinib combined with chemotherapeutic agents demonstrated synergistic killing of FLT3 mutant leukemic cells [25]. In a randomized large trial in which first relapsed patients received chemotherapy alone or followed by lestaurtinib at 80 mg twice daily, no differences were noted in terms of response and overall survival (OS) between the two arms. There was evidence of toxicity in the lestaurtinib-treated patient arm. In the lestaurtinib arm, FLT3 inhibition was highly correlated with remission rate, but target inhibition on day 15 was achieved in only 58% of cases [26].

Sorafenib (BAY-43-9006) is an oral multikinase inhibitor with activity against FLT3 and several receptor tyrosine kinases. It has been approved for the treatment of hepatocellular carcinoma and renal cell carcinoma. In contrast with other first-generation FLT3 inhibitors, it has shown activity against AML with FLT3-ITD mutation as a single-agent in the setting of post-transplant relapse and in combination with chemotherapy for newly diagnosed AML. In a phase 1 clinical trial, a clinical response was observed in 56% of cases [27]. A phase 1/2 study of sorafenib combined with idarubicin and high-dose cytarabine showed 85% of response in relapsed patients including FLT3-mutated patients highlighting the potent inhibitory effect on FLT3-mutant patients [28]. With a median follow-up of 54 weeks, the probability of survival at one year was 74%. These results were confirmed by a similar study, in which sorafenib was also given alone as maintenance therapy after consolidation [29]. In the upfront setting, addition of sorafenib to chemotherapy doubled the one-year OS in older AML patients with FLT3-ITD mutation (62% *vs* 30%; p < 0.0001) [30]. Induction chemotherapy consisted of cytarabine 100 mg/m^2 on days 1 to 7 and daunorubicin 60 mg/m^2 on days 1 to 3, with oral sorafenib 400 mg twice daily on days 1 to 7. Those not achieving a hypoplastic bone marrow on day 14 were to receive a second cycle of cytarabine and daunorubicin (5+2) plus sorafenib 400 mg twice daily for 7 days. Post-remission therapy consisted of intermediate dose cytarabine (2 g/m^2 for 5 days), with sorafenib 400 mg twice daily on days 1 to 28 for 2 cycles followed by maintenance sorafenib 400 mg twice daily for twelve 28-day cycles. The median disease-free survival (DFS) and OS were 12.5 months and 15 months respectively in patients with FLT3-ITD mutation, and 9 months and 16.2 months in patients with FLT3-TKD mutation. The 30-day induction mortality was 9%, and there were no treatment-related deaths during the phases of consolidation and maintenance. In combination with hypomethylating agents (azacitidine), sorafenib has also shown promising efficacy. The overall response rate was 46% [31] and the median OS for responders was 7.8 months. A large phase 3 study demonstrated the efficacy of sorafenib in the upfront setting of young patients with FLT3$^-$ AML in terms of event-free survival (EFS) and relapse-free survival supporting the role of protein kinase inhibitors in this patient population [32]. Sorafenib in combination with low-dose homoharringtonine [33] or low-dose cytarabine [34] also showed interesting results in primary refractory FLT3-ITD$^+$ AML. A retrospective study suggested potential benefit of post-transplant sorafenib in FLT3-ITD AML [35]. Sorafenib patients had improved 2-year progression-free survival (PFS) (82% *vs* 53%; p = 0.02) and lower 2-year cumulative incidence of relapse (8.2% *vs* 37.7%; p = 0.007) as compared to a control cohort without sorafenib. A phase 1 study investigated sorafenib as maintenance therapy after myeloablative or reduced intensity HSCT with promising results deserving a further randomized study [36]. Sorafenib after HSCT has also been suggested as a preventive strategy in high-risk patients [37].

Midostaurin (PKC412) inhibits both FLT3-ITD and FLT3-TKD kinase activity. In a phase 1 trial for refractory/relapsed AML or high-risk myelodysplastic syndrome, midostaurin at 75 mg 3 times daily decreased the peripheral blast count by 50% in most patients and the marrow blast count only in few patients [38]. These first results suggested further investigation of midostaurin in combination with other agents. Combination with all-trans retinoic acid and cladribine/cytarabine/granulocyte colony-stimulating factor chemotherapy showed 22% CR rate and 11% CR with incomplete blood recovery in relapsed/refractory patients [39]. Evaluating the efficacy of midostaurin in combination with intensive induction chemotherapy and as single agent maintenance therapy after allogeneic HSCT or high-dose cytarabine has shown efficacy in FLT3-ITD mutated AML [40]. In a large phase 3 trial, addition of midostaurin to chemotherapy showed improvements in terms of survival in younger adults, as compared to chemotherapy alone [41]. Induction therapy consisted of daunorubicin 60 mg/m^2 on days 1 to 3 and cytarabine 200 mg/m^2 on days 1 to 7. Patients were randomized to receive either orally administered midostaurin 50 mg twice daily or placebo. CR rate was not significantly different between the two arms but patients receiving midostaurin had a significantly better OS and a 5-year EFS. The median OS was 74.7 months in the midostaurin arm *versus* 26 months in the placebo arm. The median EFS was 8 months *versus* 3 months, respectively. The median time to allogeneic HSCT was similar in both arms. The benefit of addition of midostaurin to standard chemotherapy followed by one-year of midostaurin maintenance therapy was consistent across all FLT3 subgroups for both OS and EFS in both uncensored and censored for transplant analyses. No statistically significant differences were observed between midostaurin and placebo arms in terms of grade $\geq$ 3 hematologic or non-hematologic adverse events [41]. Phase 2 trials studying the efficacy of midostaurin in combination with azacitidine (NCT01093573) or decitabine (NCT01846624) in older patients are currently ongoing.

Tandutinib (MLN518) is a potent FLT3 inhibitor that also inhibits PDGFR and c-KIT. A first phase 1 study showed some response in relapsed/refractory patients [42]. The dose limiting toxicity was reversible muscle weakness.

SECOND-GENERATION FLT3 TYROSINE KINASE INHIBITOR QUIZARTINIB

The second-generation FLT3 inhibitor, quizartinib (formerly known as AC220) is both a highly selective and potent inhibitor of FLT3. Its chemical structure is $C_{29}H_{32}N_6O_4S$. Quizartinib inhibits FLT3 with low nanomolar potency in biochemical and cellular assays and is highly selective when screened against the majority of the human protein kinome, especially when compared with first-generation FLT3 inhibitors [43]. Initial studies are encouraging based on a high response rate when administered as monotherapy. However, patients with FLT3 mutations may experience greater clinical benefit when quizartinib is administered in combination with intensive chemotherapy. In addition to FLT3, quizartinib inhibits c-KIT stem cell factor CD117 which is expressed in more than 70% of AML.

Clinical Pharmacology

In a bioavailability study testing the oral administration of 60 mg of quizartinib in healthy volunteers, the median time to maximum plasma concentration (T_{max}) was 4 hours for quizartinib and 8 hours for its major active metabolite, AC886. The mean terminal half-life was 64.9 hours and 53.5 hours, respectively. Quizartinib and AC886 showed dose-proportional increases in area under the concentration *versus* time curve (AUC) and maximum plasma concentrations (C_{max}) over the tested dose range of 30 to 90 mg. Following a single-oral dose of [^{14}C]-quizartinib, 76.3% of total radioactivity was recovered in feces with only 1.6% recovered in urine 14 days after dosing. AC886 was the only major circulating metabolite and is formed by CYP3A4.

Studies with Quizartinib Used as Monotherapy

In the first-in-human phase 1 study, CP001, quizartinib was administered with intermittent dosing (14 days on drug followed by 14 days rest) from 12 mg to 450 mg, and continuous dosing at 200 mg and 300 mg for 28 days in 76 patients with relapsed/refractory AML, regardless of FLT3-ITD mutation status. An *in vitro* plasma inhibitory assay showed rapid and durable inhibition of FLT3-ITD phosphorylation as early as 2 hours after the first dose. The overall response rates were about 30% in all patients, 53% inFLT3-ITD$^+$ patients and 14%in FLT3-ITD$^-$ patients [44]. The maximum tolerated dose (MTD) was 200 mg continuous daily dosing.

Results of a phase 2 study in relapsed/refractory FLT3$^+$ AML patients demonstrated a CR rate of 51% with a median OS of 25 weeks [45]. These CR rates were confirmed in all phase 2 studies using quizartinib as a single-agent [46, 47]. However, 50% of patients relapse within 3 months. Importantly, 35% of FLT3-ITD patients were bridged to HSCT [48]. The mechanism of resistance is the development of acquired mutations in the tyrosine kinase domain of the gene.

A phase 2b study was subsequently conducted, which enrolled 76 patients with FLT3-ITD$^+$ AML randomized to 60 mg or 30 mg daily, to examine efficacy and toxicity at these lower doses. The study showed an overall response rate (ORR) not significantly different among the two doses (61% with 30 mg/day and 71% with 60 mg/day) [49]. The median OS was 20.7 weeks with 30 mg/day and 25.4 weeks with 60 mg/day. In this study, quizartinib as second salvage or post HSCT in FLT3-ITD$^+$ AML patients demonstrated efficacy with an acceptable safety profile as compared to higher doses used previously.

In phase 2 studies, patients with an allelic ratio of FLT3-ITD to total FLT3 of more than 10% were considered as ITD$^+$. However, following the results of phase

2 studies [46 - 49], the cutoff in current clinical trials with quizartinib has been reduced to $\geq 3\%$ to define quantifiable ITD mutation.

Combination Studies with Intensive Chemotherapy

AML is a polyclonal disease and patients with FLT3-ITD mutated AML may experience greater clinical benefit when quizartinib is administered in combination with standard chemotherapy. In addition to FLT3, quizartinib inhibits c-KIT stem cell factor CD117 which is expressed in more than 70% of AML cases. Furthermore, KIT mutations occur in more than 40% of core binding factor (CBF) leukemias [50].

A phase 1, open-label, multiple-dose, dose-escalation study (study 2689-CL-0005) was performed in patients with newly diagnosed AML: FLT3-ITD$^+$ or FLT3-ITD$^-$. The dose-escalation was conducted using a modified 3+3 design with 6 patients enrolled at each dose level. Patients were given a "3+7" standard induction and high-dose intermittent cytarabine for consolidation. They were allowed to proceed to HSCT after achieving a response or receive further quizartinib if they were not transplant eligible. Three dose levels were tested: 60 mg for 7 days, 60 mg for 14 days, and 40 mg for 14 days. The maximum tolerated dose was identified as 40 mg for 14 days or 60 mg for 7 days. The most common (10%) grade 3 or 4 treatment-related adverse events were febrile neutropenia (26%), thrombocytopenia (21%), anemia (21%), neutropenia (21%), leucopenia (16%), and nausea (11%) [51].

The AML 18 pilot quizartinib dose-escalation study was conducted in the United Kingdom that enrolled newly diagnosed FLT3$^+$ and FLT3$^-$ AML patients greater than 60 years of age [52]. Quizartinib was combined with standard chemotherapy comprising daunorubicin, cytarabine, and etoposide. Six cohorts with escalating doses of quizartinib (60 mg, 90 mg or 135 mf for 7 or 14 days) were planned. De-escalation to 40 mg for 7 or 14 days was allowed if 60 mg was not tolerated. Fifty-five patients (median age: 69 years) were enrolled. Death within 30 days occurred in 6.5% of evaluable patients. CR was achieved in 79%, including all FLT3-ITD$^+$ patients. Median OS at the time of analysis was 15 months.

Among current clinical trial, a phase 1/2 trial with quizartinib in combination with 5-azacitidine or low-dose cytarabine is ongoing in patients > 60 years of age with previously untreated AML (NCT01892371). An ongoing phase 3 (QuANTUM-R), open-label study of quizartinib monotherapy *versus* salvage chemotherapy in FLT3-ITD mutated patients who are refractory or relapsed in 6 months with or without HSCT is also still enrolling (NCT02039726). The QuANTUM-FIRST protocol, a phase 3, double-blind, placebo-controlled study of quizartinib administered with induction and consolidation, and administered as maintenance therapy in subjects 18 to 75 years old with newly diagnosed FLT3-ITD$^+$ AML has just opened (NCT02668653). In this study, quizartinib is administered orally at the dose of 40 mg/day for 14 days during induction and consolidation phases, and at the dose of 60 mg/day (after checking for good tolerance at 30 mg/day during the first 15 days) during the maintenance phase.

Quizartinib as Maintenance Therapy

Allogeneic HSCT represents a standard of care for consolidating FLT3-ITD AML patients in first CR [53,54]. A phase 1 study of quizartinib as maintenance therapy is ongoing but showed promising early efficacy data in AML patients who have been allografted in first or second remission. All patients were FLT3-ITD$^+$ at diagnosis, and received 40 mg/day or 60 mg/day of quizartinib after HSCT [55]. Both doses were well tolerated. The median number of cycles administered was 18.

Terminal Myeloid Differentiation

In FLT3-ITD$^+$ AML patients who responded to quizartinib therapy, the bone marrow and peripheral blood specimens displayed unexpected findings, especially in patients with normal cytogenetics. The bone marrow remained hypercellular, continued to express the FLT3-ITD mutation, and displayed progressive myeloid differentiation over time [56]. This was associated with a clinical differentiation syndrome. In *in vitro* studies, FLT3 inhibition with quizartinib induced cell-cycle arrest and differentiation rather than apoptosis. This effect might in part be mediated by the protective effects of the bone marrow microenvironment [57]. In preexisting CEBPα mutation, blastic cells failed to differentiate suggesting a mechanism of resistance to FLT3 inhibition.

Toxicity

In the phase 2 study (AC220-002), 35% of subjects experienced grade 3 QT prolongation at 200mg dose and therefore the dose was reduced. A single case of grade 4 QT prolongation was reported in a patient with pneumonia and atrial fibrillation, taking concomitant medications known to cause QT prolongation [58]. No deaths related to QT prolongation have been reported. QT prolongation was dose-dependent [49]. The incidence of drug-induced Sweet's syndrome following FLT3 inhibitor monotherapy is about 10% [59]. It is however unlikely to occur in the setting of concurrent cytoreduction with chemotherapy. The symptoms can generally be managed with corticosteroid administration. Pyodermag-angrenosum characterized by multiple cutaneous ulcers with mucopurulent or hemorrhagic exudates has also been reported [60]. It can also be managed by systemic and/or topical corticosteroids [61]. Other common adverse events observed in the phase 1 and 2 studies included gastrointestinal disorders (nausea, diarrhea, and vomiting), fatigue, and hematologic disorders (anemia, neutropenia, and thrombocytopenia). Although hematologic toxicity was associated with underlying disease, safety reports indicated delayed recovery or continued suppression of absolute neutrophil counts and platelets as a consequence of continued treatment with quizartinib.

Drug-Drug Interactions

A drug-drug interaction study assessing the effect of strong and moderate CYP3A4 inhibitors on quizartinib pharmacokinetics showed that concomitant ketoconazole, a strong CYP3A4 inhibitor, and concomitant fluconazole, a moderate CYP3A4 inhibitor, resulted in an increase in quizartinib AUC_{0-inf} values, approximately 2-fold and 1.2-fold, respectively. Additionally, concomitant ketoconazole and concomitant fluconazole resulted in an increase in predicted quizartinib C_{max} at steady state after repeat daily dosing, approximately 2-fold and 1.2-fold, respectively.

Inhibition of c-KIT

As many tyrosine kinase inhibitors, quizartinib has activity against c-KIT, a receptor tyrosine kinase which is essential for normal hematopoiesis [62]. The c-KIT receptor is an important marker of long-term hematopoietic stem cells, and it also plays an important role in hair and skin pigmentation [63]. Inhibition of c-KIT can translate into clinical significant marrow suppression, particularly when it occurs in the setting of cytotoxic chemotherapy. The more potent c-KIT inhibitors impair erythroid and myeloid progenitor cell function, but FLT3 inhibition probably has little effect on hematopoiesis (Table **2**).

Table 2. c-KIT and myelosuppressive activity of protein kinase inhibitors (according to Galanisand Levis 2015 [63]).

Protein kinase inhibitor	*In vivo* c-KIT inhibition	Hair depigmentation	Myelosuppression
Sunitinib	Yes	Yes	Yes
Quizartinib	Yes	Yes	Yes
Crenolanib	No	No	No
Sorafenib	No	No	No

PERSPECTIVES WITH OTHER SECOND- OR THIRD-GENERATION FLT3 TYROSINE KINASE INHIBITORS

Next to quizartinib, other second- or third-generation FLT3 tyrosine kinase inhibitors are under investigation. They could even be more promising in the way they could show higher cytotoxicity and overcome the emergence of therapeutic resistance that could be observed with quizartinib and first-generation FLT3 inhibitors.

Crenolanib represents a next-generation receptor tyrosine kinase that effectively suppresses growth of leukemic cells harboring both FLT3-ITD and FLT3-TKD mutations, the latter of which are increasingly seen to emerge as resistant mutations after FLT3 inhibitor therapy [64]. Crenolanib displays type I tyrosine kinase inhibitor properties and can bind the active or inactive conformations of receptor tyrosine kinases, unlike type II inhibitors which only bind the inactive form. Most inhibitors with predominantly type II properties have minimal activity against FLT3-TKD AML. This selects for the persistence of TKD clones or emergence of new ones, which in part explains the development of therapeutic resistance. Crenolanib was demonstrated to have activity against mutations in the activation loop of FLT3 [65]. In a phase 1 study, the recommended dose was 100 mg twice daily [66]. In a phase 2 clinical trial, the dose was 100 mg three times daily due to its half-life of 8 to 9 hours, which was well-tolerated and showed promising clinical activity [67]. Eleven heavily pre-treated patients (of whom 6 received prior FLT3 inhibitors) were included into a dose escalation study testing crenolanib in combination with chemotherapy [68]. Primary results showed a CR with incomplete blood count recovery in 23% of patients (including older adults; median age: 61 years), but in only 5% of patients who had received prior FLT3 inhibitors [68]. Crenolanib can be safely administered with high-dose cytarabine/idarubicin salvage in multiply relapsed AML patients [69]. Crenolanib is currently being investigated in combination with intensive chemotherapy in patients with newly diagnosed AML with a FLT3-ITD or TKD mutation (NCT02283177). First results of combination with cytarabine/anthracycline induction chemotherapy and high-dose cytarabine consolidation showed that crenolanib can be safely administered at full dose (100 mg daily) [70].

Gilteritinib (formerly known as ASP2215) is a potent inhibitor of both FLT3-ITD and FLT3-TKD mutations. In the FLT3 mutant adult patient population (median age: 61 years), the overall response rate was 57% [71]. In adult AML, a phase 1 study in combination with intensive chemotherapy is ongoing (NCT02236013), as well as a phase 3 study randomizing gilteritinib *versus* savage chemotherapy (NCT02421939). A differentiation response to gilteritinib has recently been demonstrated among relapsed/refractory FLT3-mutated patients with NPM1 and DNMT3A mutations [72].

Pacritinib (SB1518) is a third-generation tyrosine kinase inhibitor with activity against a number of targets of relevance to AML. Pacritinib is a low molecular-weight compound with potent inhibitory activities against FLT3 and JAK2 [73]. Although JAK2 mutations are rare in AML, the JAK-STAT pathway is frequently activated and may represent a mechanism of resistance to FLT3 inhibitors. Blockade of FLT3 in conjunction with JAK2 signaling could enhance clinical benefit for AML patients harboring a FLT3-ITD mutation [74]. The first clinical experience of pacritinib in AML demonstrated encouraging data in terms of tolerability and, in the challenging setting of relapsed/refractory FLT3-mutated AML, clinical response in one-third of evaluable patients including patients of more than 60 years [75]. Pacritinib was administered at an oral dose of 200 mg twice daily.

Ponatinib (AP24534), approved as a BCR-ABL inhibitor, is a multi-kinase inhibitor. It is a potent type I FLT3 inhibitor with activity against FLT3 with induced point mutations [76, 77]. Preclinical data has demonstrated potent activity against FLT3-mutated AML cell lines as well as in animal models [78]. In a phase 1 trial, in heavily pre-treated AML patients including FLT3-ITD mutants, the overall response was 25% [79]. The most commonly noted toxicity was pancreatitis. Ponatinibis not currently that much developed as a FLT3 inhibitor due to increased risk of vascular adverse events. While ponatinib has demonstrated activity in tyrosine kinase inhibitor-resistant chronic myeloid leukemia, irrespective of BCR-ABL kinase domain mutation, it has been tested against clinically relevant FLT3-ITD mutant isoforms that confer resistance to quizartinib or sorafenib [76]. Substitution of the FLT3 gatekeeper phenylalanine with leucine (F691L) conferred mild resistance to ponatinib, but substitutions at the FLT3 activation loop residue D835 conferred a high degree of resistance. Alternative strategies should therefore be required for patients with tyrosine kinase inhibitor-resistant FLT3-ITD D835 mutations.

CONCLUSIONS

AML is particularly challenging because it displays a large variety of molecular abnormalities, even in the malignant cell burden of a single individual. This genetic diversity allows the leukemia or components of the leukemia to escape from the cytotoxic effects of targeted drugs. Several FLT3 inhibitors have been evaluated in clinical trials, either as single agents or in combination with chemotherapy. However, most candidates either did not generate sufficient initial response or failed to sustain therapeutic benefit, primarily due to development of secondary resistance. The most promising areas of research are therefore the elucidation of the mechanisms of resistance to FLT3 inhibitors. The primary cause of resistance is the acquisition of point mutations in the ATP-binding region of the FLT3-TKD, altering the conformational state and weakening the binding affinity to specific FLT3 inhibitors. With first-generation FLT3 inhibitors, peripheral blasts decline but bone marrow responses were very low. First-generation FLT3 inhibitors, including lestaurtinib and midostaurin, did not have optimal potency, specificity or pharmacokinetic properties. Sorafenib is an effective FLT3 inhibitor, but its activity is often lost over time. Point mutations in the FLT3-ITD kinase domain were identified as a mediator of resistance to midostaurin [80] and mutations in TKD1 or TKD2 as a mediator of resistance to sorafenib [81]. Secondary TKD mutation in FLT3-ITD patients treated with an FLT3 inhibitor represents another cause of resistance [82]. New agents can overcome FLT3 inhibitor resistance. The introduction of potent and targeted agents such as quizartinib for FLT3-mutant disease has increased hope for this frequently letal subtype of AML. At present, quizartinib appears as the most potent and specific FLT3 inhibitor and is able to completely suppress FLT3-ITD autophosphorylation. The combination of its excellent potency, selectivity, and pharmacokinetic properties made it the first drug candidate with a profile that matches the characteristics desirable for a clinical FLT3 inhibitor [43]. However, like sorafenib, resistance can appear after short periods of treatment with acquired mutations in FLT3-ITD that disrupt binding of the drug to the target [83, 84]. Substitutions at the gatekeeper residues such as FLT3-ITD (F691) are well documented causes of resistance to kinase inhibitors. Analogues of the FLT3-ITD (D835V) activation loop mutation have also proven problematic for a number of kinase inhibitors. Quizartinib and sorafenib are type II FLT3 inhibitors that bind to the inactive conformation of the kinase and prevent its activation. Substitutions at F691 and D835 in FLT3-ITD pose substantial barriers to disease control in patients treated with quizartinib or sorafenib. In contrast, type I inhibitors target the active conformation of the kinase and may be effective against FLT3-ITD with point mutations conferring resistance to type II FLT3 inhibitors. Crenolanib could then represent an even more promising FLT3 inhibitor. It has activity against mutations in the activation loop of FLT3 [65]. In this way, it could represent a pan-selective FLT3 of choice that could overcome quizartinib resistance. Tested against a panel of D835 mutant cell lines, it showed superior cytotoxicity compared with quizartinib [85]. Combining type I and type II tyrosine kinase inhibitors could also be help enhance efficacy and perhaps suppress the emergence of therapeutic resistance. Little is known about mechanisms of resistance to crenolanib [65]. The existence of independent alternative survival pathways for malignant cells either through further genetic lesions or metabolic adaptation, were other potential mechanisms for failures. Simultaneous targeting of independent pathways will render leukemic cells less likely to escape FLT3 mono-inhibition. mTOR signaling is downstream of FLT3 kinase and is important for leukemia cell survival, suggesting that combination of mTOR and FLT3 inhibitors may have potential for clinical activity [86]. Targeting JAK2, which is the case when using pacritinib, provides here an interesting opportunity.

Because patients with FLT3-ITD AML often relapse during consolidation therapy, it has recently been suggested that these patients did not benefit from intensifying therapy, in contrast to those with FLT3 wild type [87]. Three distinct genotypes have been identified: normal $FLT3^{WT/WT}$, heterozygous $FLT3^{ITD/WT}$, and hemizygous $FLT3^{ITD/-}$, FLT3-ITD$^+$ corresponding to the lack of FLT3 WT allele [88]. $FLT3^{ITD/-}$ patients showed a significantly shortened survival when compared with $FLT3^{WT/WT}$ or $FLT3^{ITD/WT}$ patients, demonstrating a greater growth advantage conferred by this genotype. FLT3 ligand counteracts the effects of FLT3 inhibition by directly interacting with the mutated receptor. Rapid and sustained increases in FLT3 ligand levels associated with induction and consolidation chemotherapy have been described in several clinical trials [26, 89]. The sequence of administration of FLT3 inhibitors when combined with chemotherapy might therefore be of great concern. Leukemia stem cells harboring the FLT3-ITD mutation could have a survival advantage over their wild-type counterparts and emerge at relapse as a dominant clone. It has therefore been suggested that the best therapeutic approach would be to proceed as rapidly as possible to allogeneic HSCT once remission is achieved. FLT3 mutated AML is a disease that evolves from diagnosis to relapse. In FLT3-ITD mutated AML, cells at relapse appear to be more addicted to FLT3 signaling compared to cells at the time of initial diagnosis. One might therefore argue that for newly diagnosed FLT3-ITD mutated AML, a combination with a FLT3 inhibitor that carries relatively broad kinase activity should be used, whereas in the relapsed setting a more selective FLT3 inhibitor might be indicated.

FLT3 inhibitors, especially currently quizartinib, should represent one of many novel therapeutic approaches as a step forward for patients with FLT3-ITD mutations. Combination of chemotherapy with FLT3 inhibitors continues to look a promising approach since results from large combination trials appear to show a survival advantage. A better understanding of the pathobiology of underlying mechanisms of resistance to FLT3 inhibitors should overcome or even prevent the occurrence of mutations responsible for treatment failure. In this setting, first results with next-generation tyrosine kinase inhibitors appear very encouraging.

CONSENT FOR PUBLICATION

Not applicable.

CONFLICT OF INTEREST

The authors confirm that this chapter content has no conflict of interest.

ACKNOWLEDGEMENT

Declared none.

REFERENCES

[1]   Guidelines NCCN. (version 1.2015) www.nccn.org/professionals/physician_gls/pdf/aml.pdf

[2]   Ohtake S, Miyawaki S, Fujita H, *et al.* Randomized study of induction therapy comparing standard-dose idarubicin with high-dose daunorubicin in adult patients with previously untreated acute myeloid leukemia: the JALSG AML201 Study. Blood 2011; 117(8): 2358-65.
[http://dx.doi.org/10.1182/blood-2010-03-273243] [PMID: 20693429]

[3]   Burnett AK, Russell NH, Hills RK, *et al.* A randomized comparison of daunorubicin 90 mg/m^2*vs* 60 mg/m^2 in AML induction: results from the UK NCRI AML17 trial in 1206 patients. Blood 2015; 125(25): 3878-85.
[http://dx.doi.org/10.1182/blood-2015-01-623447] [PMID: 25833957]

[4]   Burnett AK, Russell NH, Hills RK, *et al.* Optimization of chemotherapy for younger patients with acute myeloid leukemia: results of the medical research council AML15 trial. J Clin Oncol 2013; 31(27): 3360-8.
[http://dx.doi.org/10.1200/JCO.2012.47.4874] [PMID: 23940227]

[5]   Fernandez HF, Sun Z, Yao X, *et al.* Anthracycline dose intensification in acute myeloid leukemia. N Engl J Med 2009; 361(13): 1249-59.
[http://dx.doi.org/10.1056/NEJMoa0904544] [PMID: 19776406]

[6]   Shurin MR, Esche C, Lotze MT. FLT3: receptor and ligand. Biology and potential clinical application. Cytokine Growth Factor Rev 1998; 9(1): 37-48.
[http://dx.doi.org/10.1016/S1359-6101(97)00035-X] [PMID: 9720755]

[7]   Drexler HG. Expression of FLT3 receptor and response to FLT3 ligand by leukemic cells. Leukemia 1996; 10(4): 588-99.
[PMID: 8618433]

[8]   Gilliland DG, Griffin JD. The roles of FLT3 in hematopoiesis and leukemia. Blood 2002; 100(5): 1532-42.
[http://dx.doi.org/10.1182/blood-2002-02-0492] [PMID: 12176867]

[9]   Levis M, Small D. FLT3 tyrosine kinase inhibitors. Int J Hematol 2005; 82(2): 100-7.
[http://dx.doi.org/10.1532/IJH97.05079] [PMID: 16146839]

[10]   Nakao M, Yokota S, Iwai T, *et al.* Internal tandem duplication of the flt3 gene found in acute myeloid leukemia. Leukemia 1996; 10(12): 1911-8.
[PMID: 8946930]

[11]   Abu-Duhier FM, Goodeve AC, Wilson GA, Care RS, Peake IR, Reilly JT. Identification of novel FLT-3 Asp835 mutations in adult acute myeloid leukaemia. Br J Haematol 2001; 113(4): 983-8.
[http://dx.doi.org/10.1046/j.1365-2141.2001.02850.x] [PMID: 11442493]

[12]   Mizuki M, Fenski R, Halfter H, *et al.* Flt3 mutations from patients with acute myeloid leukemia induce transformation of 32D cells mediated by the Ras and STAT5 pathways. Blood 2000; 96(12): 3907-14.
[PMID: 11090077]

[13]   Thiede C, Steudel C, Mohr B, *et al.* Analysis of FLT3-activating mutations in 979 patients with acute myelogenous leukemia: association with FAB subtypes and identification of subgroups with poor prognosis. Blood 2002; 99(12): 4326-35.
[http://dx.doi.org/10.1182/blood.V99.12.4326] [PMID: 12036858]

[14]   Fröhling S, Schlenk RF, Breitruck J, *et al.* Prognostic significance of activating FLT3 mutations in younger adults (16 to 60 years) with acute myeloid leukemia and normal cytogenetics: a study of the

AML Study Group Ulm. Blood 2002; 100(13): 4372-80.
[http://dx.doi.org/10.1182/blood-2002-05-1440] [PMID: 12393388]

[15]　Whitman SP, Maharry K, Radmacher MD, *et al.* FLT3 internal tandem duplication associates with adverse outcome and gene- and microRNA-expression signatures in patients 60 years of age or older with primary cytogenetically normal acute myeloid leukemia: a Cancer and Leukemia Group B study. Blood 2010; 116(18): 3622-6.
[http://dx.doi.org/10.1182/blood-2010-05-283648] [PMID: 20656931]

[16]　Schlenk RF, Kayser S, Bullinger L, *et al.* Differential impact of allelic ratio and insertion site in FLT3-ITD-positive AML with respect to allogeneic transplantation. Blood 2014; 124(23): 3441-9.
[http://dx.doi.org/10.1182/blood-2014-05-578070] [PMID: 25270908]

[17]　Ossenkoppele G, Schuurhuis GJ. MRD in AML: does it already guide therapy decision-making? Hematology 2016. Education Program of the American Society of Hematology 2016; pp. 356-65.

[18]　Döhner H, Estey EH, Amadori S, *et al.* Diagnosis and management of acute myeloid leukemia in adults: recommendations from an international expert panel, on behalf of the European LeukemiaNet. Blood 2010; 115(3): 453-74.
[http://dx.doi.org/10.1182/blood-2009-07-235358] [PMID: 19880497]

[19]　Fiedler W, Serve H, Döhner H, *et al.* A phase 1 study of SU11248 in the treatment of patients with refractory or resistant acute myeloid leukemia (AML) or not amenable to conventional therapy for the disease. Blood 2005; 105(3): 986-93.
[http://dx.doi.org/10.1182/blood-2004-05-1846] [PMID: 15459012]

[20]　Fiedler W, Kayser S, Kebenko M, *et al.* Sunitinib and intensive chemotherapy in patients with acute myeloid leukemia and activating FLT3 mutations: results of the AMLSG 10-07 study. Blood 2012; 120(21): 1483.

[21]　Levis M, Allebach J, Tse KF, *et al.* A FLT3-targeted tyrosine kinase inhibitor is cytotoxic to leukemia cells *in vitro* and *in vivo*. Blood 2002; 99(11): 3885-91.
[http://dx.doi.org/10.1182/blood.V99.11.3885] [PMID: 12010785]

[22]　Smith BD, Levis M, Beran M, *et al.* Single-agent CEP-701, a novel FLT3 inhibitor, shows biologic and clinical activity in patients with relapsed or refractory acute myeloid leukemia. Blood 2004; 103(10): 3669-76.
[http://dx.doi.org/10.1182/blood-2003-11-3775] [PMID: 14726387]

[23]　Marshall JL, Kindler H, Deeken J, *et al.* Phase I trial of orally administered CEP-701, a novel neurotrophin receptor-linked tyrosine kinase inhibitor. Invest New Drugs 2005; 23(1): 31-7.
[http://dx.doi.org/10.1023/B:DRUG.0000047103.64335.b0] [PMID: 15528978]

[24]　Knapper S, Burnett AK, Littlewood T, *et al.* A phase 2 trial of the FLT3 inhibitor lestaurtinib (CEP701) as first-line treatment for older patients with acute myeloid leukemia not considered fit for intensive chemotherapy. Blood 2006; 108(10): 3262-70.
[http://dx.doi.org/10.1182/blood-2006-04-015560] [PMID: 16857985]

[25]　Levis M, Pham R, Smith BD, Small D. *In vitro* studies of a FLT3 inhibitor combined with chemotherapy: sequence of administration is important to achieve synergistic cytotoxic effects. Blood 2004; 104(4): 1145-50.
[http://dx.doi.org/10.1182/blood-2004-01-0388] [PMID: 15126317]

[26]　Levis M, Ravandi F, Wang ES, *et al.* Results from a randomized trial of salvage chemotherapy followed by lestaurtinib for patients with FLT3 mutant AML in first relapse. Blood 2011; 117(12): 3294-301.
[http://dx.doi.org/10.1182/blood-2010-08-301796] [PMID: 21270442]

[27]　Zhang W, Konopleva M, Shi YX, *et al.* Mutant FLT3: a direct target of sorafenib in acute myelogenous leukemia. J Natl Cancer Inst 2008; 100(3): 184-98.
[http://dx.doi.org/10.1093/jnci/djm328] [PMID: 18230792]

[28]   Ravandi F, Cortes JE, Jones D, *et al.* Phase I/II study of combination therapy with sorafenib, idarubicin, and cytarabine in younger patients with acute myeloid leukemia. J Clin Oncol 2010; 28(11): 1856-62.
[http://dx.doi.org/10.1200/JCO.2009.25.4888] [PMID: 20212254]

[29]   Al-Kali A, Cortes J, Faderl S, *et al.* Patterns of molecular response to and relapse after combination of sorafenib, idarubicin, and cytarabine in patients with FLT3 mutant acute myeloid leukemia. Clin Lymphoma Myeloma Leuk 2011; 11(4): 361-6.
[http://dx.doi.org/10.1016/j.clml.2011.06.007] [PMID: 21816375]

[30]   Uy GL, Mandrekar S, Laumann K, *et al.* Addition of sorafenib to chemotherapy improves the overall survival of older adults with FLT3-ITD mutated acute myeloid leukemia (AML) (Alliance C11001). Blood 2015; 126(23): 319.
[PMID: 25852056]

[31]   Ravandi F, Alattar ML, Grunwald MR, *et al.* Phase 2 study of azacytidine plus sorafenib in patients with acute myeloid leukemia and FLT-3 internal tandem duplication mutation. Blood 2013; 121(23): 4655-62.
[http://dx.doi.org/10.1182/blood-2013-01-480228] [PMID: 23613521]

[32]   Röllig C, Müller-Tidow C, Hüttmann A, *et al.* Sorafenib *versus* placebo in addition to standard therapy in younger patients with newly diagnosed acute myeloid leukemia: results from 267 patients treated in the randomized placebo-controlled SAL-SORAML trial. Blood 2014; 124(21): 6.

[33]   Xu G, Mao L, Liu H, Yang M, Jin J, Qian W. Sorafenib in combination with low-dos--homoharringtonine as a salvage therapy in primary refractory FLT3-ITD-positive AML: a case report and review of literature. Int J Clin Exp Med 2015; 8(11): 19891-4.
[PMID: 26884901]

[34]   Liu XS, Long H, Huang YX, *et al.* Clinical efficacy of sorafenib combined with low dose cytarabine for treating patients with FLT3+ relapsed and refractory acute myeloid leukemia. Zhongguo Shi Yan Xue Ye Xue Za Zhi 2016; 24(2): 394-8.
[PMID: 27150998]

[35]   Brunner AM, Li S, Fathi AT, *et al.* Haematopoietic cell transplantation with and without sorafenib maintenance for patients with FLT3-ITD acute myeloid leukaemia in first complete remission. Br J Haematol 2016; 175(3): 496-504.
[http://dx.doi.org/10.1111/bjh.14260] [PMID: 27434660]

[36]   Chen YB, Shuli L, Andrew LA, *et al.* Phase I trial of maintenance sorafenib after allogeneic hematopoietic stem cell transplantation for patients with FLT3-ITD AML. Blood 2015; 124(23): 671.

[37]   Sharma M, Ravandi F, Bayraktar UD, *et al.* Treatment of FLT3-ITD-positive acute myeloid leukemia relapsing after allogeneic stem cell transplantation with sorafenib. Biol Blood Marrow Transplant 2011; 17(12): 1874-7.
[http://dx.doi.org/10.1016/j.bbmt.2011.07.011] [PMID: 21767516]

[38]   Stone RM, DeAngelo DJ, Klimek V, *et al.* Patients with acute myeloid leukemia and an activating mutation in FLT3 respond to a small-molecule FLT3 tyrosine kinase inhibitor, PKC412. Blood 2005; 105(1): 54-60.
[http://dx.doi.org/10.1182/blood-2004-03-0891] [PMID: 15345597]

[39]   Ramsingh G, Westervelt P, McBride A, *et al.* Phase I study of cladribine, cytarabine, granulocyte colony stimulating factor (CLAG regimen) and midostaurin and all-trans retinoic acid in relapsed/refractory AML. Int J Hematol 2014; 99(3): 272-8.
[http://dx.doi.org/10.1007/s12185-014-1503-4] [PMID: 24488798]

[40]   Schlenk R, Döhner K, Salih H, *et al.* Midostaurin in combination with intensive induction and as single agent maintenance therapy after consolidation therapy with allogeneic hematopoietic stem cell transplantation or high-dose cytarabine (NCT01477606). Blood 2015; 126(23): 322.

[41] Stone RM, Mandrekar S, Sanford BL, *et al.* The multi-kinase inhibitor midostaurin (M) prolongs survival compared with placebo (P) in combination with daunorubicin (D)/cytarabine (C) induction (ind), high-dose C consolidation (consol), and as maintenance (maint) therapy in newly diagnosed acute myeloid leukemia (AML) patients (pts) age 18-60 with FLT3 mutations (muts): an international prospective randomized (rand) P-controlled double-blind trial (CALGB 10603/RATIFY [Alliance]). Blood 2015; 126(23): 6.
[PMID: 26138538]

[42] DeAngelo DJ, Stone RM, Heaney ML, *et al.* Phase 1 clinical results with tandutinib (MLN518), a novel FLT3 antagonist, in patients with acute myelogenous leukemia or high-risk myelodysplastic syndrome: safety, pharmacokinetics, and pharmacodynamics. Blood 2006; 108(12): 3674-81.
[http://dx.doi.org/10.1182/blood-2006-02-005702] [PMID: 16902153]

[43] Zarrinkar PP, Gunawardane RN, Cramer MD, *et al.* AC220 is a uniquely potent and selective inhibitor of FLT3 for the treatment of acute myeloid leukemia (AML). Blood 2009; 114(14): 2984-92.
[http://dx.doi.org/10.1182/blood-2009-05-222034] [PMID: 19654408]

[44] Cortes JE, Kantarjian H, Foran JM, *et al.* Phase I study of quizartinib administered daily to patients with relapsed or refractory acute myeloid leukemia irrespective of FMS-like tyrosine kinase 3-internal tandem duplication status. J Clin Oncol 2013; 31(29): 3681-7.
[http://dx.doi.org/10.1200/JCO.2013.48.8783] [PMID: 24002496]

[45] Martinelli G, Perl AE, Dombret H, *et al.* Effect of quizartinib (AC220) on response rates and long-term survival in elderly patients with FLT3-ITD positive or negative relapsed/refractory acute myeloid leukemia. J Clin Oncol 2013; 31 (Suppl.): 7021.

[46] Cortes J, Perl A, Dombret H, *et al.* Final results of a phase 2 open-label, monotherapy efficacy and safety study of quizartinib (AC220) in patients 60 years of age with FLT3 ITD positive or negative relapsed/refractory acute myeloid leukemia. Blood 2012; 120(21): 48.

[47] Levis MJ, Perl AE, Dombret H, *et al.* Final results of a phase 2 open-label, monotherapy, efficacy and safety study of quizartinib (AC220) in patients with FLT3-ITD positive or negative relapsed/refractory acute myeloid leukemia after second line chemotherapy or hematopoietic stem cell transplantation. Blood 2012; 120(21): 673.

[48] Cortes JE, Perl AE, Dombret H, *et al.* Response rate and bridging to hematopoietic stem cell transplantation (HSCT) with quizartinib (AC220) in patients with FLT3-ITD positive or negative relapsed?refractory AML after second-line chemotherapy or previous bone marrow transplant. J Clin Oncol 2013; 31 (Suppl.): 7012.

[49] Russell N, Tallman MS, Goldberg S, *et al.* Quizartinib (AC220) in patients with FLT3-ITD(+) relapsed or refractory acute myeloid leukemia: final results of a randomized phase 2 study. Haematologica 2014; 99 (Suppl. 1): 333.

[50] Cairoli R, Grillo G, Beghini A, *et al.* C-Kit point mutations in core binding factor leukemias: correlation with white blood cell count and the white blood cell index. Leukemia 2003; 17(2): 471-2.
[http://dx.doi.org/10.1038/sj.leu.2402795] [PMID: 12592353]

[51] Altman JK, Foran JM, Pratz KW, *et al.* Results of a phase I study of quizartinib (AC220, ASP2689) in combination with induction and consolidation chemotherapy in younger patients with newly diagnosed acute myeloid leukemia. Blood 2013; 122(21): 623.
[PMID: 23908441]

[52] Burnett AK, Bowen D, Russell N, *et al.* AC220 (quizartinib) can be safely combined with conventional chemotherapy in older patients with newly diagnosed acute myeloid leukaemia:Experience from the AML 18 pilot study. Blood 2013; 122(21): 622.

[53] Brunet S, Martino R, Sierra J. Hematopoietic transplantation for acute myeloid leukemia with internal tandem duplication of FLT3 gene (FLT3/ITD). Curr Opin Oncol 2013; 25(2): 195-204.
[http://dx.doi.org/10.1097/CCO.0b013e32835ec91f] [PMID: 23385863]

[54]   Stone RM. Acute myeloid leukemia in first remission: to choose transplantation or not? J Clin Oncol 2013; 31(10): 1262-6.
[http://dx.doi.org/10.1200/JCO.2012.43.4258] [PMID: 23439752]

[55]   Sandmaier BM, Khaled SK, Oran B, *et al.* Results of a phase 1 study of quizartinib (AC220) as maintenance therapy in subjects with acute myeloid leukemia in remission following allogeneic hematopoietic cell transplantation. Blood 2014; 124(21): 428.

[56]   Sexauer A, Perl A, Yang X, *et al.* Terminal myeloid differentiation *in vivo* is induced by FLT3 inhibition in FLT3/ITD AML. Blood 2012; 120(20): 4205-14.
[http://dx.doi.org/10.1182/blood-2012-01-402545] [PMID: 23012328]

[57]   Yang X, Sexauer A, Levis M. Bone marrow stroma-mediated resistance to FLT3 inhibitors in FLT3-ITD AML is mediated by persistent activation of extracellular regulated kinase. Br J Haematol 2014; 164(1): 61-72.
[http://dx.doi.org/10.1111/bjh.12599] [PMID: 24116827]

[58]   Levis MJ, Cortes JE, Perl EA, *et al.* FLT3 inhibitor-induced neutrophilicdermatosis Blood 2013; 122(2): 239-42.

[59]   Fathi AT, Le L, Hasserjian RP, Sadrzadeh H, Levis M, Chen YB. FLT3 inhibitor-induced neutrophilic dermatosis. Blood 2013; 122(2): 239-42.
[http://dx.doi.org/10.1182/blood-2013-01-478172] [PMID: 23687091]

[60]   Jha PK, Rana A, Kapoor S, Kher V. Pyoderma gangrenosum in a renal transplant recipient: A case report and review of literature. Indian J Nephrol 2015; 25(5): 297-9.
[http://dx.doi.org/10.4103/0971-4065.156900] [PMID: 26628796]

[61]   Callen JP. Pyoderma gangrenosum. Lancet 1998; 351(9102): 581-5.
[http://dx.doi.org/10.1016/S0140-6736(97)10187-8] [PMID: 9492798]

[62]   Lyman SD, Jacobsen SE. c-kit ligand and Flt3 ligand: stem/progenitor cell factors with overlapping yet distinct activities. Blood 1998; 91(4): 1101-34.
[PMID: 9454740]

[63]   Galanis A, Levis M. Inhibition of c-Kit by tyrosine kinase inhibitors. Haematologica 2015; 100(3): e77-9.
[http://dx.doi.org/10.3324/haematol.2014.117028] [PMID: 25425690]

[64]   Zimmerman EI, Turner DC, Buaboonnam J, *et al.* Crenolanib is active against models of drug-resistant FLT3-ITD-positive acute myeloid leukemia. Blood 2013; 122(22): 3607-15.
[http://dx.doi.org/10.1182/blood-2013-07-513044] [PMID: 24046014]

[65]   Smith CC, Lasater EA, Lin KC, *et al.* Crenolanib is a selective type I pan-FLT3 inhibitor. Proc Natl Acad Sci USA 2014; 111(14): 5319-24.
[http://dx.doi.org/10.1073/pnas.1320661111] [PMID: 24623852]

[66]   Lewis NL, Lewis LD, Eder JP, *et al.* Phase I study of the safety, tolerability, and pharmacokinetics of oral CP-868,596, a highly specific platelet-derived growth factor receptor tyrosine kinase inhibitor in patients with advanced cancers. J Clin Oncol 2009; 27(31): 5262-9.
[http://dx.doi.org/10.1200/JCO.2009.21.8487] [PMID: 19738123]

[67]   Collins R, Kantarjian HM, Levis MJ, *et al.* Clinical activity of crenolanib in patients with D835 mutant FLT3-positive relapsed/refractory acute myeloid leukemia (AML). J Clin Oncol 2014; 32 (Suppl.): 5s.
[http://dx.doi.org/10.1200/jco.2014.32.15_suppl.7027]

[68]   Randhawa JK, Kantarjian HM, Borthakur G, *et al.* Results of a phase II study of crenolanib in relapsed/refractory acute myeloid leukemia patients (Pts) with activating FLT3 mutations. Blood 2014; 124(21): 389.

[69]   Cortes J, Ravandi F, Garcia-Manero G, *et al.* Dose escalation study of crenolanibin combination with high dose cytarabine/idarubicin salvage chemotherapy in multiply relapsed FLT3 positive AML.

Haematologica 2016; 101(S1): 377.

[70] Wang E, Stone R, Tallman M, *et al.* Safety study of crenolanib, a type I FLT3 inhibitor, with cytarabine/daunorubicin or cytarabine/idarubicin induction and high-dose cytarabine consolidation in newly diagnosed FLT3⁺AML. Haematologica 2016; 101(S1): 41.

[71] Levis MJ, Perl AE, Altman JK, *et al.* Results of a first-in-human, phase I/II trial of ASP2215, a selective, potent inhibitor of FLT3/AXL in patients with relapsed or refractory (R/R) acute myeloid leukemia (AML). J Clin Oncol 2015; 33 (Suppl.): 7003.

[72] Canaani J, Rea B, Sargent R, *et al.* Differentiation response to gilteritinib (ASP2215) in relapsed/refractory FLT3 mutated acute myeloid leukemia patients is associated with co-mutations in NPM1 and DNMT3A. Haematologica 2016; 101(S1): 42.

[73] William AD, Lee AC, Blanchard S, *et al.* Discovery of the macrocycle 11-(2-pyrrolidin-1-yl-et-oxy)-14,19-dioxa-5,7,26-triaza-tetracyclo[19.3.1.1(2,6).1(8,12)]hept-cosa-1(25),2(26),3,5,8,10,12(27),16,21,23-decaene (SB1518), a potent Janus kinase 2/fms-like tyrosine kinase-3 (JAK2/FLT3) inhibitor for the treatment of myelofibrosis and lymphoma. J Med Chem 2011; 54(13): 4638-58.
[http://dx.doi.org/10.1021/jm200326p] [PMID: 21604762]

[74] Hart S, Goh KC, Novotny-Diermayr V, *et al.* Pacritinib (SB1518), a JAK2/FLT3 inhibitor for the treatment of acute myeloid leukemia. Blood Cancer J 2011; 1(11): e44.
[http://dx.doi.org/10.1038/bcj.2011.43] [PMID: 22829080]

[75] Knapper S, Grech A, Cahalin P, *et al.* An evaluation of the tyrosine kinase inhibitor pacritinib in patients with relapsed FLT3-mutated acute myeloid leukaemia (the UK NCRI AML17 study). Haematologica 2016; 101(S1): 40.

[76] Zirm E, Spies-Weisshart B, Heidel F, *et al.* Ponatinib may overcome resistance of FLT3-ITD harbouring additional point mutations, notably the previously refractory F691I mutation. Br J Haematol 2012; 157(4): 483-92.
[http://dx.doi.org/10.1111/j.1365-2141.2012.09085.x] [PMID: 22409268]

[77] Smith CC, Lasater EA, Zhu X, *et al.* Activity of ponatinib against clinically-relevant AC220-resistant kinase domain mutants of FLT3-ITD. Blood 2013; 121(16): 3165-71.
[http://dx.doi.org/10.1182/blood-2012-07-442871] [PMID: 23430109]

[78] Gozgit JM, Wong MJ, Wardwell S, *et al.* Potent activity of ponatinib (AP24534) in models of FLT3-driven acute myeloid leukemia and other hematologic malignancies. Mol Cancer Ther 2011; 10(6): 1028-35.
[http://dx.doi.org/10.1158/1535-7163.MCT-10-1044] [PMID: 21482694]

[79] Shah NP, Talpaz M, Deininger MW, *et al.* Ponatinib in patients with refractory acute myeloid leukaemia: findings from a phase 1 study. Br J Haematol 2013; 162(4): 548-52.
[http://dx.doi.org/10.1111/bjh.12382] [PMID: 23691988]

[80] Heidel F, Solem FK, Breitenbuecher F, *et al.* Clinical resistance to the kinase inhibitor PKC412 in acute myeloid leukemia by mutation of Asn-676 in the FLT3 tyrosine kinase domain. Blood 2006; 107(1): 293-300.
[http://dx.doi.org/10.1182/blood-2005-06-2469] [PMID: 16150941]

[81] Zhang W, Gao C, Konopleva M, *et al.* Reversal of acquired drug resistance in FLT3-mutated acute myeloid leukemia cells *via* distinct drug combination strategies. Clin Cancer Res 2014; 20(9): 2363-74.
[http://dx.doi.org/10.1158/1078-0432.CCR-13-2052] [PMID: 24619500]

[82] Alvarado Y, Kantarjian HM, Luthra R, *et al.* Treatment with FLT3 inhibitor in patients with FLT3-mutated acute myeloid leukemia is associated with development of secondary FLT3-tyrosine kinase domain mutations. Cancer 2014; 120(14): 2142-9.
[http://dx.doi.org/10.1002/cncr.28705] [PMID: 24737502]

[83]   Smith CC, Wang Q, Chin CS, *et al.* Validation of ITD mutations in FLT3 as a therapeutic target in human acute myeloid leukaemia. Nature 2012; 485(7397): 260-3.
[http://dx.doi.org/10.1038/nature11016] [PMID: 22504184]

[84]   Pauwels D, Sweron B, Cools J. The N676D and G697R mutations in the kinase domain of FLT3 confer resistance to the inhibitor AC220. Haematologica 2012; 97(11): 1773-4.
[http://dx.doi.org/10.3324/haematol.2012.069781] [PMID: 22875611]

[85]   Galanis A, Ma H, Rajkhowa T, *et al.* Crenolanib is a potent inhibitor of FLT3 with activity against resistance-conferring point mutants. Blood 2014; 123(1): 94-100.
[http://dx.doi.org/10.1182/blood-2013-10-529313] [PMID: 24227820]

[86]   Chen W, Drakos E, Grammatikakis I, *et al.* mTOR signaling is activated by FLT3 kinase and promotes survival of FLT3-mutated acute myeloid leukemia cells. Mol Cancer 2010; 9: 292.
[http://dx.doi.org/10.1186/1476-4598-9-292] [PMID: 21067588]

[87]   Levis M. FLT3/ITD AML and the law of unintended consequences. Blood 2011; 117(26): 6987-90.
[http://dx.doi.org/10.1182/blood-2011-03-340273] [PMID: 21586749]

[88]   Whitman SP, Archer KJ, Feng L, *et al.* Absence of the wild-type allele predicts poor prognosis in adult de novo acute myeloid leukemia with normal cytogenetics and the internal tandem duplication of FLT3: a cancer and leukemia group B study. Cancer Res 2001; 61(19): 7233-9.
[PMID: 11585760]

[89]   Serve H, Krug U, Wagner R, *et al.* Sorafenib in combination with intensive chemotherapy in elderly patients with acute myeloid leukemia: results from a randomized, placebo-controlled trial. J Clin Oncol 2013; 31(25): 3110-8.
[http://dx.doi.org/10.1200/JCO.2012.46.4990] [PMID: 23897964]

CHAPTER 2

Physical Exercise for Cancer Patients Treated with Chemotherapy

Shinichiro Morishita[1,*], Atsuhiro Tsubaki[1] and **Jack B. Fu[2]**

[1] *Institute for Human Movement and Medical Sciences, Niigata University of Health and Welfare, Niigata, Japan*

[2] *Department of Palliative, Rehabilitation & Integrative Medicine, University of Texas MD Anderson Cancer Center, Houston, TX, USA*

Abstract: The use of chemotherapy in the treatment of cancer, although increasingly efficacious for improving survival, produces short- and long-term negative physiological side effects. Sleep disturbance, fatigue, and depressed mood are common and distressing problems that occur during and after chemotherapy. Furthermore, after chemotherapy, cancer patients tend to experience decreased cardiorespiratory fitness and muscle strength. These changes lead to a decrease in physical function and quality of life (QoL). Physical exercise has been shown to improve physical function and QoL in cancer patients during and after chemotherapy. Physical exercise may also alleviate symptoms that interfere with physical fatigue, mental fatigue, treatment-related fatigue, muscle pain, arthralgia, and other pain, depression, anxiety, and insomnia. Furthermore, physical exercise prolongs survival and reduces the mortality of these patients. Based on previous scientific reports, this chapter introduces the role of physical exercise in the care of cancer patients treated with chemotherapy.

Keywords: Cancer, Exercise, Oncology, Physical Function, Physiotherapy, Rehabilitation.

INTRODUCTION

Cancer patients have some physical and psychological side effects during and after cancer treatment [1, 2]. Many cancer patients who have received chemotherapy are at risk of developing long-term side effects such as sleep disturbance, fatigue, and depressed mood, which are common and distressing problems that occur during and after chemotherapy [3, 4]. Additionally, cancer patients tend to have sedentary and decreased physical activity following

* **Corresponding author Shinichiro Morishita:** Institute for Human Movement and Medical Sciences, Niigata University of Health and Welfare, Niigata, Japan, Shimami-cho 1398, Kita-ku, Niigata City 950-3198, Japan; Tel: +81252574300; Fax: +81252574300; E-mail: ptmorishin@yahoo.co.jp

Atta-ur-Rahman (Ed.)

chemotherapy [5]. Physical and psychological impairments can also lead to substantial social problems, such as the inability to work or fulfill other normative social roles [6].

Physical exercise is important for cancer patients to improve their physical function and physical activity [7, 8] (Fig. **1**). It can help alleviate many common side effects [9] and improve cancer patients' functioning and reduce their symptom levels [10]. Some reports showed the effect of physical exercise on patients with lung [11, 12], ovarian [13], breast [14, 15], gastrointestinal [16], and colorectal cancers [17], and leukemia [18]. A systematic review reported that physical exercise improves physical function in patients with cancer following chemotherapy [19]. Moreover, physical exercise is effective for reducing cancer-related fatigue during and after cancer treatment and significantly better than the available pharmaceutical options [20]. Furthermore, physical exercise may also alleviate symptoms that interfere with daily life of cancer patients and survivors, such as appetite loss, diarrhea, paresthesia, constipation, physical fatigue, mental fatigue, treatment related fatigue, muscle pain, arthralgia and other pain, depression, anxiety and insomnia [21]. Hence, physical exercise has shown to provide benefits to patients both during and after cancer treatment.

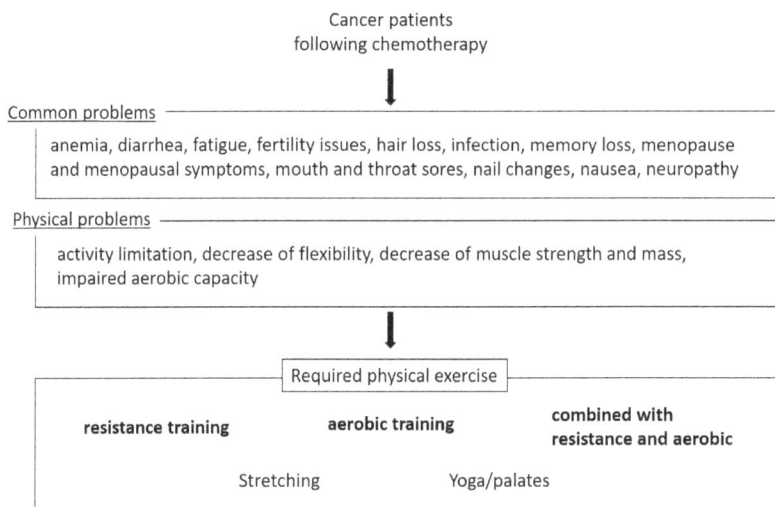

Cancer patients
following chemotherapy

↓

Common problems

anemia, diarrhea, fatigue, fertility issues, hair loss, infection, memory loss, menopause and menopausal symptoms, mouth and throat sores, nail changes, nausea, neuropathy

Physical problems

activity limitation, decrease of flexibility, decrease of muscle strength and mass, impaired aerobic capacity

↓

Required physical exercise

resistance training   aerobic training   combined with resistance and aerobic

Stretching   Yoga/palates

Fig. (1). Cancer patients have some problems following chemotherapy and required physical exercise.

A literature review was conducted using PubMed and Scopus to select studies involving physical exercise and various cancer patients treated with chemotherapy. The review included observational studies to investigate physical, psychological, and psychosocial side effects during and after chemotherapy.

Additionally, the review included interventional studies to investigate the effect of physical exercise on cancer patients treated with chemotherapy. This chapter presented the effect of physical exercise on patients with different cancer types undergoing chemotherapy.

Breast Cancer

Physical, Psychological, and Psychosocial Side Effects During and After Chemotherapy

Several observational studies have investigated physical, psychological, and psychosocial side effects during and after chemotherapy in breast cancer patients [22, 23]. Chemotherapy-induced peripheral neuropathy (CIPN) is a neuropathic disorder caused by neurotoxic chemotherapeutic agents including taxanes [24], which are frequently administered in breast cancer. Previous studies showed that some side effects include anemia/low red blood cell count, diarrhea, fatigue, fertility issues, hair changes, infection, memory loss, menopause and menopausal symptoms, mouth and throat sores, nail changes, nausea, neuropathy (problems with hands and feet), taste and smell changes, vaginal dryness, vomiting, weight changes, bone loss/osteoporosis, heart problems, and vision/eye problems [22, 23].

Decreased Physical Function During and After Chemotherapy

Some observational studies have investigated physical function during and after chemotherapy in breast cancer patients [25 - 32]. Breast cancer patients have often experienced decreased physical function [25, 26], such as upper extremity and trunk dysfunctions, fatigue, weight gain, pain, chemotherapy☐induced peripheral neuropathy, and activity and performance limitations [27]; moreover, physical function became a prognostic biomarker among breast cancer survivors [28]. Breast cancer patients have decreased muscle strength following chemotherapy [29] and showed markedly impaired shoulder and knee strengths and joint dysfunctions before and after anticancer treatment [30]. Moreover, aerobic capacity and muscle strength were significantly decreased following chemotherapy [31]. Breast cancer patients sometime experienced lymphedema with chemotherapy, but not with axillary node surgery [32].

Type of Physical Exercise (Aerobic, Resistance Training, or Combined, Flexibility)

Many studies have shown the effect of aerobic, resistance training, or their combination, and flexibility exercises include stretching, foam rolling, yoga, tai chi, and Pilates [33 - 36] (Table **1**).

A meta-analysis found that aerobic exercise may improve cancer-related fatigue in breast cancer patients receiving chemotherapy [37]. Aerobic exercise is often used to improve the exercise capacity in such patients [38]. Interestingly, aerobic exercise improves their QoL [39]. Aerobic exercises with moderate or higher

Table 1. Recommendation type physical exercise on different cancer diagnosis.

Type	Decreased Physical Function	Recommendation Type of Exercise
Breast	Aerobic capacity	Aerobic exercise
	Muscle strength / mass	Resistance training,
	Flexibility	Combined
	Lymphedema	Stretching, yoga, tai chi, and Pilates
Prostate	Aerobic capacity	Aerobic exercise
	Muscle strength / mass	Resistance training,
		Combined
Lung	Aerobic capacity	Aerobic exercise
	Muscle strength / mass	Resistance training,
		Combined
Colon and Rectum	Not applicable	Aerobic exercise
		Resistance training,
		Combined
Stomach (Gastric)	Not applicable	Not applicable
Esophagus	Not applicable	Not applicable
Lymphoma, leukemia (Adult)	Aerobic capacity	Aerobic exercise
	Muscle strength / mass	Resistance training,
		Combined
Lymphoma, leukemia (Child)	Aerobic capacity	Aerobic exercise
	Muscle strength / mass	Resistance training,
		Combined
Gynecologic	Not applicable	Combined
Others	Not applicable	Not applicable

intensities are safe exercise prescriptions in breast cancer patients [40]. A previous study showed that a 10-week moderate-intensity aerobic exercise program significantly improves QoL and physical functioning of breast cancer survivors [41]. Moderate-to-high intensity aerobic training when conducted with one-on-one supervised program is a safe adjunct therapy, which is associated with improvements in cardiopulmonary function during neoadjuvant chemotherapy [42].

Resistance training has similar effects on muscle strength upper and lower extremity [43] and bone density [44]. Moreover, resistance training is effective in improving the QoL and physical function [45]. Resistance exercises consist of supervised, group-based exercise program (2/week over 12 weeks) that could improve fatigue for breast cancer patients undergoing chemotherapy [46]. Resistance training consisted of 2 days/week of 10 exercises including two sets of 8–12 repetitions at 60% of the participants' one-repetition maximum [34]. Another study showed that breast cancer patients completed a high load (6-8 repetition maximum) and low load (15-20 repetition maximum) exercise session consisting of two sets of five upper body resistance exercises [47]. Another study used resistance training which consisted of two days/week of 10 exercises including two sets of 8–12 repetitions at 52%–69% of their 1 repetition maximum [45]. Resistance training was not decided based on the set, repetition, intensity, and duration for breast cancer patients, but most studies used the guidelines of the American College of Sports Medicine [48].

Most studies used aerobic and resistance combined exercise for breast cancer patients [49]. In a systematic review, aerobic and resistance training combined exercise has effectiveness on fatigue, cardiovascular fitness, and muscle strength and slight effect on cancer-related fatigue, health-related QoL, and depression [50, 51] and aerobic and resistance training combined exercise influences the metabolic syndrome of breast cancer patients [52, 53]. Combined resistance and aerobic training comprise an 8-week exercise program with 3 weekly sessions of 90-min duration [53]. This study showed the improvement of the QoL and overall physical fitness of women breast cancer survivors [53].

Furthermore, there are other exercises for breast cancer patients. Yoga during chemotherapy resulted in modest short-term improvements in sleep quality [36]. Yoga and aerobic exercise have an impact on fatigue in breast cancer patients [54]. Another study investigated the effects of a yoga program on reducing psychological symptoms in breast cancer patients receiving chemotherapy [55].

Pilates is sometimes used to decrease lymphedema in these patients [56]. A systematic review revealed that Pilates improves the range of motion, pain, and fatigue in breast cancer patients [57].

Prostate

Physical, Psychological, and Psychosocial Side Effects During and After Chemotherapy

There are some observational studies that have investigated physical, psychological, and psychosocial side effects during and after chemotherapy in prostate cancer patients [58 - 60]. Some side effects of chemotherapy are hair loss, mouth sores, appetite loss, nausea and vomiting, diarrhea, and fatigue [58, 59]. A systematic review found that the prevalence of depression and anxiety in men with prostate cancer is relatively high [60].

Decreased Physical Function During and After Chemotherapy

Several observational studies have investigated physical function during and after chemotherapy in prostate cancer patients [61 - 65]. Prostate cancer patients slightly showed functional declined on Instrumental Activities of Daily Living (IADL), whereas physical function remained stable and QoL improved during chemotherapy [61]. In prostate cancer, some studies investigated the relationship between physical function and androgen deprivation therapy (ADT) [62 - 64]. ADT has a significant effect on body composition, walking speed, and physical performance in prostate cancer patients [62, 63]. ADT is associated with declines in self-reported physical functioning and upper body muscle strength as well as worse lower body muscle strength relative to prostate cancer controls [64]. Patients experienced unintentional weight loss (4.5 kg in past year), exhaustion (self-reported), weakness (grip strength, lowest 20%), slow walking speed (15 feet, slowest 20%), and low physical activity (kcal/week, lowest 20%) [65].

Type of Physical Exercise (Aerobic, Resistance Training, or Combined, Flexibility)

Many studies have shown the effects of physical exercise in prostate cancer patients [66 - 74]. Physical exercise offers some benefits to men undergoing treatment for prostate cancer [66]. In a systematic review, prostate cancer patients reported the benefits and effects of physical exercise on body composition, physical fitness, functional performance, QoL, and fatigue [67]. Another systematic review showed that physical exercise can influence cancer-related fatigue [68]. Many studies using resistance training or resistance training plus aerobic exercise have been conducted. However, to date, no study has investigated the effect of aerobic exercise for prostate cancer. Resistance training is often used in prostate cancer patients following ADT; the 12-week resistance training intervention effectively improved sarcopenia, body fat %, strength, and QoL in prostate cancer patients [69]. Another study also reported that resistance training improved upper and lower muscle strength in prostate cancer patients [70]. Similarly, resistance training has effects on QoL as well as muscle strength in these patients [69]. Thus, resistance training is effective in improving the physical function and QoL of prostate cancer patients. Some studies used combined aerobic and resistance training exercises for prostate cancer patients [71]. Combined physical exercises can affect muscle strength and exercise tolerance in prostate cancer patients following ADT [72]. Another study showed that aerobic and resistance exercises appear to have beneficial effects on QoL among prostate cancer patients [73, 74]. In prostate cancer patients with ADT therapy, many studies evaluated the effectiveness of resistance training or aerobic exercise. However, no study has investigated the effect of physical exercise on prostate cancer patients undergoing chemotherapy.

Lung

Physical, Psychological, and Psychosocial Side Effects During and After Chemotherapy

There are some observational studies that have investigated physical, psychological, and psychosocial side effects during and after chemotherapy in lung cancer patients [75 - 77]. Breathlessness is a common symptom in lung cancer and other cancer-related illnesses with systemic anticancer treatment such as chemotherapy [75]. Lung cancer patients experience some side effects such as hair loss, nausea, vomiting, mouth sores, appetite loss, exhaustion, easy bruising or bleeding, and increased risk of infection because of destruction of white blood cells [76, 77].

Decreased Physical Function During and After Chemotherapy

There are some observational studies that have investigated physical function during and after chemotherapy in lung cancer patients [78 - 82]. Lung cancer patients have decreased skeletal muscle depletion, muscle strength, and walking capacity [78]. A study showed that geriatric patients with lung cancer have decreased physical function symptom severity, which varied according to the treatment type, stage of disease, and sex [79]. Lung cancer patients have often decreased QoL as well as decreased physical function following chemotherapy [80]. The physical function of patients with advanced lung cancer was poorer with increasing fatigue [81]. This study also showed that Karnofsky Performance Status, weakness, and Hospital Anxiety and Depression Scale scores were correlated independently with fatigue [81]. Shortness of breath is a common problem for people with metastatic non-small cell lung cancer. Lung cancer patients often experience dyspnea [82]. Chemotherapy to the chest may damage the lungs. However, lung cancer patients have been shown to experience dyspnea because of smoking (current/former), not chemotherapy.

Type of Physical Exercise (Aerobic, Resistance Training, or Combined, Flexibility)

Many studies have shown the effects of physical exercise in lung cancer patients [83 - 89]. A systematic review showed that lung cancer patients may benefit from physical exercise both during and after treatment [83]. Physical exercises improve the exercise capacity and health-related QoL of patients with lung cancer [84]. Physical exercise is an important facet of health care management in lung cancer patients [85]. Physical exercises consist of resistance-, cardiovascular-, and relaxation training 4 h weekly for 6 weeks and a concurrent unsupervised home-based exercise program [86]. Aerobic exercise was carried out for lung cancer patients to improve exercise capacity and muscle strength [87]. Another study conducted an 8-week high-intensity aerobic interval training for lung cancer patients [88]. In addition, another study showed that progressive resistance exercise training in post-treatment lung cancer survivor showed significant improvements in muscle strength, endurance capacity, and QoL [89]. Some studies showed the effect of combined resistance and aerobic exercise for lung cancer patients [86]. An 8-week combined rehabilitation program in lung cancer patients significantly improved pulmonary function and functional and peak exercise capacity in contrast to baseline [90]. In inoperable lung cancer patients undergoing chemotherapy, combined exercises significantly improve muscle strength and exercise capacity [12].

Colon and Rectum

Physical, Psychological, and Psychosocial Side Effects During and After Chemotherapy

To date, there are few observational studies that have investigated physical, psychological, and psychosocial side effects during and after chemotherapy in colon and rectal cancer patients [91]. Some side effects encountered by colon and rectum cancer patients are hair loss, mouth sores, appetite loss, nausea and vomiting, diarrhea, hand-foot syndrome, neuropathy, allergic or sensitivity reactions, diarrhea, neutropenia, polyneuropathy, bruising or bleeding, and fatigue [91].

Decreased Physical Function During and After Chemotherapy

To date, limited studies have shown the physical function of colon and rectum cancer patients. In elderly patients with colorectal cancer, the Eastern Cooperative

Oncology Group performance status 2 group had more limitations in IADLs, lower baseline QoL, and a lower patient-rated health score [92].

Type of Physical Exercise (Aerobic, Resistance Training, or Combined, Flexibility)

Some studies have shown the effects of physical exercise in colon and rectal cancer patients [93 - 98]. Aerobic exercise reduces visceral adipose tissue in a dose-response fashion among patients with stage I-III colon cancer [93]. Aerobic exercise is feasible in patients with colorectal cancer after completion of adjuvant chemotherapy [94]. Similarly, rectal cancer patients undergoing neoadjuvant chemoradiotherapy have improved physical functioning, cardiovascular endurance, and QoL after supervised aerobic exercise [95]. Compared to control, an 18-week supervised aerobic and muscle strength training significantly decreased fatigue and improved exercise capacity in colon cancer patients compared [96]. Another study investigated the recruitment rate of exercise programs for colon cancer patients undergoing adjuvant chemotherapy [97]. Colon cancer patients with combined resistance and aerobic exercise were able to maintain or improve their physical fitness levels and maintain or decrease their fatigue levels during chemotherapy [97]. A study reported that colon cancer survivors reported the effects of physical exercise on physical function and QoL [98]. Overall, reports on physical exercise for colon and rectum cancer patients in contrast to that in breast or lung cancer patients are limited.

Stomach (Gastric)

Physical, Psychological, and Psychosocial Side Effects During and After Chemotherapy

To date, there are few observational studies that have investigated physical, psychological, and psychosocial side effects during and after chemotherapy in stomach cancer patients [99, 100]. Most stomach cancer patients experienced some side effects such as nausea and vomiting, appetite loss, hair loss, diarrhea, mouth sores, increased risk of infection (low white blood cell count), bleeding, or bruising after minor cuts or injuries (low platelet count) [99, 100].

Decreased Physical Function During and After Chemotherapy

Previous reports have shown that chemotherapy induced weakness and skeletal muscle fatigue [101]. Stomach cancer patients tend to have decreased physical function after chemotherapy. However, to date, there no reports investigating this decrease during and after chemotherapy.

Type of Physical Exercise (Aerobic, Resistance Training, or Combined, Flexibility)

Stomach cancer patients are likely to have decreased physical function, muscle strength, and aerobic capacity following chemotherapy; however, there are no reports on the decreased physical function during and after chemotherapy. Furthermore, to date, there are no recommended physical exercises for stomach cancer patients following chemotherapy.

Esophagus

Physical, Psychological, and Psychosocial Side Effects During and After Chemotherapy

To date, there are few studies that have investigated physical, psychological, and psychosocial side effects during and after chemotherapy in esophageal cancer patients [99, 102]. Esophageal cancer patients undergoing chemotherapy have significant oral mucositis associated with regimen-related toxicities [103]. Esophagus cancer patients also reported bone marrow suppression, nausea and vomiting, diarrhea, hair loss, mouth sore, fatigue, appetite loss, skin changes, constipation, flu-like symptoms, nervous system problems, and pain following chemotherapy [99, 102].

Decreased Physical Function During and After Chemotherapy

Some reports have shown decreased physical function in esophageal cancer patients during and after chemotherapy [104 - 106]. Esophageal and gastric cancers can cause significant physical decline [104]. Neoadjuvant chemotherapy also has a negative effect on cardiopulmonary physical fitness in esophageal cancer patients [105]. Previous reports have shown that reduced fitness and physical functioning are long-term sequelae after curative treatment for esophageal cancer [106].

Type of Physical Exercise (Aerobic, Resistance Training, or Combined, Flexibility)

Esophageal cancer patients are likely to have decreased physical function following chemotherapy; however, there are no reports on decreased physical function during and after chemotherapy. Furthermore, to date, there are no recommended physical exercises for esophagus cancer patients following chemotherapy.

Lymphoma, Leukemia (Adult)

Physical, Psychological, and Psychosocial Side Effects During and After Chemotherapy

There are many studies that have investigated physical, psychological, and psychosocial side effects during and after chemotherapy in lymphoma and leukemia patients [107, 108]. Lymphoma and leukemia patients experience some side effects, including hair loss, mouth sores, appetite loss, nausea and vomiting, diarrhea, increased risk of infections, easy bruising or bleeding (due to low blood platelet count), fatigue (due to low red blood cell count) following chemotherapy [107, 108].

Decreased Physical Function During and After Chemotherapy

There are some observational studies that have investigated physical function during and after chemotherapy for leukemia and lymphoma cancer patients [109 - 114]. Leukemia and lymphoma patients have experienced some decreased muscle strength and exercise capacity [109, 110]. Elderly leukemia patients have more decreased physical function in contrast to young patients [111]. On the contrary, a study showed that compared to younger adults, older leukemia adults can tolerate intensive chemotherapy quite well from QoL and physical function perspectives [112]. Furthermore, patients with leukemia and lymphoma after receiving hematopoietic stem cell transplantation experience muscle weakness, decreased endurance capacity, and fatigue in contrast to before the treatment [113, 114].

Type of Physical Exercise (Aerobic, Resistance Training, or Combined, Flexibility)

Many studies have shown the effect of physical exercise for leukemia and lymphoma patients following chemotherapy [18]. Physical exercise is a safe, promising intervention for improving fitness and QOL among these patients [18]. Most physical exercise studies have used combined aerobic and resistance training for patients with hematological malignancy. However, only a few studies used aerobic exercise for hematological malignancy patients [109]. A study showed that 3 weeks of systematic walking exercise is clinically feasible for acute myeloid leukemia patients undergoing chemotherapy and can effectively improve their fatigue-related experiences [109]. Combined exercises including aerobic and resistance training were more frequently performed by patients with hematological malignancy following chemotherapy [115]. Combined physical exercises will provide evidence of their effects on functional and physical capacity, symptom burden, and QoL in patients with acute leukemia during outpatient management [116]. Supervised combined exercise program (4–5 days per week, 30–60 min per session) significantly improved aerobic fitness, lower body strength, and grip strength in contrast to the control group in patients with acute myeloid leukemia undergoing induction chemotherapy [117]. Combined physical exercise has beneficial effects on physical performance, physical functioning, and treatment-related symptoms in cancer patients while receiving myeloablative chemotherapy [118]. A 4-week mixed-modality combined supervised exercise in adults with acute leukemia undergoing induction chemotherapy improved fatigue and timed up and go test scores in contrast to the control group [119]. In leukemia patients, combined aerobic and strength training exercises administered 3 times per week, twice daily, for 30 min have significant improvements in cardiorespiratory endurance and reductions in total fatigue scores after intervention [120].

Lymphoma, Leukemia (Child and Adolescent)

Physical, Psychological, and Psychosocial Side Effects During and After Chemotherapy

There are many studies that have investigated physical, psychological, and psychosocial side effects during and after chemotherapy in pediatric lymphoma and leukemia patients [121, 122]. Children and adolescent with hematological malignancies experienced some side effects such as hair loss, mouth sores, appetite loss, diarrhea, anorexia-cachexia syndrome, nausea and vomiting, increased risk of infections (due to low white blood cell count), easy bruising and bleeding (from low platelet count), and fatigue (due to low red blood cell count) following chemotherapy [121]. Furthermore, children and adolescents with hematological malignancies reported different side effects [122]. They also have growth, development, hormone, learning and memory, and heart-related problems [122].

Decreased Physical Function During and After Chemotherapy

Some observational studies have investigated physical function during and after chemotherapy in pediatric lymphoma and leukemia patients [123 - 127]. Pediatric leukemia patients have decreased physical function as well as cardiac damage, obesity, endocrine dysfunction, metabolic abnormalities, reduced muscle strength, poor sensation, and impaired balance [123]. Childhood leukemia following treatment has long-term and late cognitive, physical, and psychological effects [124, 125]. They showed a significantly lower mean strength of the knee extensors in contrast to that in control [126]. The anaerobic capacity was both significantly reduced than the reference values [126]. Long-term adult survivors of childhood acute lymphoblastic leukemia (ALL) have significantly impaired exercise capacity and left ventricular function than controls [127].

Type of Physical Exercise (Aerobic, Resistance Training, or Combined, Flexibility)

Some studies have shown the effects of physical exercise in pediatric lymphoma and leukemia patients [128 - 131]. In a systematic review, physical exercises have positive effects on fatigue, strength, and QoL, immune system, body composition, sleep, activity levels, and various aspects of physical functioning [128]. Similarly, in childhood ALL, physical exercises have some positive effects on physical fitness in the intervention group in contrast to that in the control group in terms of body composition, flexibility, cardiorespiratory fitness, muscle strength, and health-related QoL [129]. A previous report showed the effects of aerobic exercise or combined exercise [130, 131]. A 6-week home-based aerobic exercise intervention has significantly lower "general fatigue" subscale than those in the control group of children with ALL [130]. Combined aerobic and resistance training significantly improved knee extensor muscle strength in contrast to control group after a 4-month intervention [131].

Gynecologic Cancer as Cervical, Endometrial Cancer, Ovarian

Physical, Psychological, and Psychosocial Side Effects During and After Chemotherapy

There are some studies that have investigated physical, psychological, and psychosocial side effects during and after chemotherapy in gynecologic cancer patients [132 - 134]. Gynecologic cancer patients reported that following side effects: nausea, appetite loss, mouth sores, increased risk of infection, bleeding or bruising easily, vomiting, hair loss, diarrhea, peripheral neuropathy, fatigue, and sexuality and intimacy issues following chemotherapy [132, 133]. Furthermore, this side effects can diminish their QoL during and after treatment [134].

Decreased Physical Function During and After Chemotherapy

To date, there are limited studies that have investigated physical function in gynecologic cancer. Cervical, endometrial, and ovarian cancer patients may have decreased muscle strength and aerobic capacity. Thus, physical exercises are recommended to gynecologic cancer patients.

Type of Physical Exercise (Aerobic, Resistance Training, or Combined, Flexibility)

Some studies reported the effect of physical exercise on gynecologic cancer patients [13, 135 - 137]. Exercise program consist of a 12-week combined exercise with aerobic, resistance, core stability, and balance exercise, showing significant improvements in QoL, fatigue, mental health, muscular strength, and balance function after the intervention [13]. Another study showed the effect of pelvic floor muscle training for urinary incontinence among gynecologic cancer survivors [135]. This intervention significantly improved cancer survivor's urinary incontinence [135]. Walking intervention throughout chemotherapy for ovarian cancer demonstrated improvements in physical function, physical symptoms, physical well-being, and ovarian cancer-specific QoL [136]. In patients with ovarian cancer, home-based exercise and cognitive behavioral therapy have significantly improve fatigue and symptom of depression in contrast to control [137].

Others (Liver, Pancreas, Gallbladder, Bile Duct, Urinary Bladder, Skin, Oral, Thyroid Gland)

Physical, Psychological, and Psychosocial Side Effects During and After Chemotherapy

To date, limited studies have investigated physical, psychological, and psychosocial side effects during and after chemotherapy in liver, pancreas, gallbladder, bile duct, urinary bladder, skin, oral, and thyroid gland cancers. Chemotherapy for other cancers such as liver, pancreas, gallbladder, bile duct, urinary bladder, skin, oral, thyroid gland also have some side effects as chest pain, constipation, diarrhea, dyspnea, fatigue, mucositis, pain, rash, and vomiting [138].

Decreased Physical Function During and After Chemotherapy

Patients treated with neoadjuvant chemotherapy experienced greater losses in skeletal muscle area and skeletal muscle mass in patients with foregut cancer [139]. In general, treatment of cancer with chemotherapy decreases endurance capacity and muscle strength [140]. However, there are few studies that have investigated decreased physical function during and after chemotherapy in patients with liver, pancreas, gallbladder, bile duct, urinary bladder, skin, oral, or thyroid gland cancers.

Type of Physical Exercise (Aerobic, Resistance Training, or Combined, Flexibility)

Patients with these cancer types also may have decreased physical function following chemotherapy; however, there are no reports on the decreased physical function during and after chemotherapy. Furthermore, to date, there are no recommended physical exercises for these cancer patients following chemotherapy.

Managing Risk Using Physical Exercise for Cancer Patients Treated with Chemotherapy

Previous reports showed that the physical exercise is safe for patients with hematological malignancies undergoing high-dose chemotherapy as a standard therapeutic modality [141]. In general, physical exercise is contraindicated if hemoglobin is below 8 g/dl: clinicians should monitor signs and symptoms of fatigue and exertion, and adjust exercise intensity or duration to accommodate fatigue or weakness. With hemoglobin concentrations >10 g/dl, aerobic and resistance exercise are recommended as tolerated [142]. Hematological values where absolute neutrophil count less than $0.5 \times 10^9 \mu l$ and platelet count less than 50×10^9 µl is contraindicated for exercise as well [21]. Cancer patients taking immune suppressants should avoid public gyms until their white blood cell count return to safe level ($>500/mm^3$) to prevent infection [21].

A previous report showed that severely thrombocytopenic hematologic cancer patients have increased bleeding rates due to lower platelet counts [143]. However, with exercise guidelines for severely thrombocytopenic patients, the risk of severe exercise-related bleeding events was low [143]. Similarly, in children undergoing hematopoietic stem cell transplant (HSCT) during a period of severe thrombocytopenia, bleeding complications during or after mobilization and supervised exercise during physiotherapy and occupational therapy are minor and relatively rare [144]. A study showed that hematological malignancy patients receiving high-dose chemotherapy have cytopenias; hence, physical therapists wore masks, aprons, and plastic gloves to patients from acquiring infection during physical exercise [145]. In this study, there was no significant difference in complications during cytopenia between the physical therapy and control groups [145]. Even if cancer patients received immune suppressants during chemotherapy, physical exercise did not aggravate patients' condition as risk factors were properly managed.

Recommendation for Physical Exercise

Cancer survivors are recommended to engage in regular exercises consistent with the recommendations by the American College of Sports Medicine [146, 147]. Cancer patients should avoid sedentary lifestyle and actively exercise at home (Fig. **2**). A goal of 150 min of moderate-intensity exercise or 75 min of vigorous-intensity exercise or an equivalent combination aerobic exercise over 3–5 days and resistance training at least 2 days per week is recommended. Aerobic activity should be performed in episodes of at least 10 min and should be done throughout the week. When older adults cannot performed 150 min of moderate-intensity aerobic activity in a week because of chronic conditions, they should be as physically active as their abilities and conditions allow [148]. Resistance sessions should involve major muscle groups 2–3 days per week (8–10 muscle groups, 8–10 repetitions, 2 sets). Two to three resistance exercise (*e.g.*, lifting weights) sessions each week involving moderate- to vigorous-intensity exercises targeting the major muscle groups are recommended.

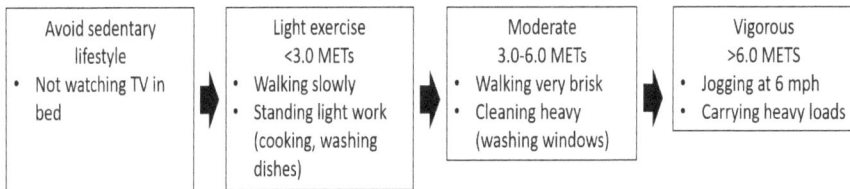

Avoid sedentary lifestyle	Light exercise <3.0 METs	Moderate 3.0-6.0 METs	Vigorous >6.0 METS
• Not watching TV in bed	• Walking slowly • Standing light work (cooking, washing dishes)	• Walking very brisk • Cleaning heavy (washing windows)	• Jogging at 6 mph • Carrying heavy loads

Fig. (2). Recommendation physical exercise for cancer patients in home.

Physical Exercise Prolongs Survival and Reduces Mortality

Recently, physical exercises have resulted in improvements in the survival and mortality of cancer patients [149]. In adult survivors of childhood cancer, increased exercise over an 8-year period was associated with a 40% reduction in all-cause mortality rate compared with maintenance of low exercise. Vigorous exercise in early adulthood and increased exercise over 8 years was associated with lower risk of mortality in adult survivors of childhood cancer [150]. In cancer patients, associations between the total duration of exercise per week and cancer mortality were found, with the lowest risk being observed at low-t--medium levels of exercise, which may minimize cancer mortality risk. With breast cancer, exercise intervention delivered during and beyond treatment for breast cancer, which was designed to cater for all women regardless of place of residence and access to health services, has clear potential to benefit survival [151]. Similarly, in breast cancer patients, exercise and physical activity above eight METs (Metabolic equivalents) h/week in contrast to lower levels of activity was significantly associated with improved overall survival [152]. In non-metastatic colorectal cancer patients, increasing levels of exercise after disease reduced cancer-specific mortality [153]. Physical exercise also improves survival and decreases mortality among patients with breast and colon cancer [149]. Exercise can significantly improve survival and reduce cancer recurrence [153], and it may also reduce the risk of further developing cancer among patients who are undergoing or have completed chemotherapy.

CONCLUSION

In conclusion, many studies have reported that physical exercise can reverse treatment-related side effects, improve muscle mass, endurance capacity, self-esteem, and QoL, and importantly slow the disease's progression and improve chances of survival. Physical exercise, as an additional therapy for cancer patients undergoing chemotherapy, is associated with many benefits. Engaging in physical exercise increases muscle strength, joint flexibility, and general condition, all of which may be impaired by chemotherapy. Future research is needed on the use of physical exercise for other cancer patients treated with chemotherapy, excluding breast, lung, leukemia, gynecologic, and prostate cancers.

ABBREVIATIONS

ADT   Androgen Deprivation Therapy

ALL   Acute Lymphoblastic Leukemia

IADL Instrumental Activities of Daily Living

QoL   Quality of Life

CONSENT FOR PUBLICATION

Not applicable.

CONFLICT OF INTEREST

The authors have no other relevant affiliations or financial involvement with any organization or entity with a financial interest in or financial conflict with the subject matter or materials discussed in the manuscript apart from those disclosed.

ACKNOWLEDGEMENTS

The authors would like to thank Editage (www.editage.jp) for the English language review.

REFERENCES

[1]   Miller KD, Siegel RL, Lin CC, *et al.* Cancer treatment and survivorship statistics, 2016. CA Cancer J Clin 2016; 66(4): 271-89.
      [http://dx.doi.org/10.3322/caac.21349] [PMID: 27253694]

[2]   Timilshina N, Breunis H, Tomlinson GA, *et al.* Long-term recovery of quality of life and physical function over three years in adult survivors of acute myeloid leukemia after intensive chemotherapy. Leukemia 2018.
      [PMID: 29884902]

[3]   Ray M, Rogers LQ, Trammell RA, Toth LA. Fatigue and sleep during cancer and chemotherapy: translational rodent models. Comp Med 2008; 58(3): 234-45.
      [PMID: 18589865]

[4]   Jim HS, Small B, Faul LA, Franzen J, Apte S, Jacobsen PB. Fatigue, depression, sleep, and activity during chemotherapy: daily and intraday variation and relationships among symptom changes. Ann Behav Med 2011; 42(3): 321-33.
      [http://dx.doi.org/10.1007/s12160-011-9294-9] [PMID: 21785899]

[5]   Fassier P, Zelek L, Partula V, *et al.* Variations of physical activity and sedentary behavior between before and after cancer diagnosis: Results from the prospective population-based NutriNet-Santé cohort. Medicine (Baltimore) 2016; 95(40): e4629.

[http://dx.doi.org/10.1097/MD.0000000000004629] [PMID: 27749527]

[6]   Dooley LN, Slavich GM, Moreno PI, Bower JE. Strength through adversity: Moderate lifetime stress exposure is associated with psychological resilience in breast cancer survivors. Stress Health 2017; 33(5): 549-57.
[http://dx.doi.org/10.1002/smi.2739] [PMID: 28052491]

[7]   McClellan R. Exercise programs for patients with cancer improve physical functioning and quality of life. J Physiother 2013; 59(1): 57.
[http://dx.doi.org/10.1016/S1836-9553(13)70150-4] [PMID: 23419919]

[8]   Piscione PJ, Bouffet E, Timmons B, *et al.* Exercise training improves physical function and fitness in long-term paediatric brain tumour survivors treated with cranial irradiation. Eur J Cancer 2017; 80: 63-72.
[http://dx.doi.org/10.1016/j.ejca.2017.04.020] [PMID: 28551430]

[9]   Gardner JR, Livingston PM, Fraser SF. Effects of exercise on treatment-related adverse effects for patients with prostate cancer receiving androgen-deprivation therapy: a systematic review. J Clin Oncol 2014; 32(4): 335-46.
[http://dx.doi.org/10.1200/JCO.2013.49.5523] [PMID: 24344218]

[10]   Oldervoll LM, Loge JH, Lydersen S, *et al.* Physical exercise for cancer patients with advanced disease: a randomized controlled trial. Oncologist 2011; 16(11): 1649-57.
[http://dx.doi.org/10.1634/theoncologist.2011-0133] [PMID: 21948693]

[11]   Quist M, Adamsen L, Rørth M, Laursen JH, Christensen KB, Langer SW. The Impact of a Multidimensional Exercise Intervention on Physical and Functional Capacity, Anxiety, and Depression in Patients With Advanced-Stage Lung Cancer Undergoing Chemotherapy. Integr Cancer Ther 2015; 14(4): 341-9.
[http://dx.doi.org/10.1177/1534735415572887] [PMID: 25800229]

[12]   Quist M, Rørth M, Langer S, *et al.* Safety and feasibility of a combined exercise intervention for inoperable lung cancer patients undergoing chemotherapy: a pilot study. Lung Cancer 2012; 75(2): 203-8.
[http://dx.doi.org/10.1016/j.lungcan.2011.07.006] [PMID: 21816503]

[13]   Mizrahi D, Broderick C, Friedlander M, *et al.* An Exercise Intervention During Chemotherapy for Women With Recurrent Ovarian Cancer: A Feasibility Study. Int J Gynecol Cancer 2015; 25(6): 985-92.
[http://dx.doi.org/10.1097/IGC.0000000000000460] [PMID: 25914961]

[14]   Courneya KS, Segal RJ, Gelmon K, *et al.* Predictors of adherence to different types and doses of supervised exercise during breast cancer chemotherapy. Int J Behav Nutr Phys Act 2014; 11: 85.
[http://dx.doi.org/10.1186/s12966-014-0085-0] [PMID: 24997476]

[15]   Courneya KS, McKenzie DC, Mackey JR, *et al.* Subgroup effects in a randomised trial of different types and doses of exercise during breast cancer chemotherapy. Br J Cancer 2014; 111(9): 1718-25.
[http://dx.doi.org/10.1038/bjc.2014.466] [PMID: 25144625]

[16]   Jensen W, Baumann FT, Stein A, *et al.* Exercise training in patients with advanced gastrointestinal cancer undergoing palliative chemotherapy: a pilot study. Support Care Cancer 2014; 22(7): 1797-806.
[PMID: 24531742]

[17]   Zimmer P, Trebing S, Timmers-Trebing U, *et al.* Eight-week, multimodal exercise counteracts a progress of chemotherapy-induced peripheral neuropathy and improves balance and strength in metastasized colorectal cancer patients: a randomized controlled trial. Support Care Cancer 2018; 26(2): 615-24.
[http://dx.doi.org/10.1007/s00520-017-3875-5] [PMID: 28963591]

[18]   Alibhai SM, O'Neill S, Fisher-Schlombs K, *et al.* A clinical trial of supervised exercise for adult inpatients with acute myeloid leukemia (AML) undergoing induction chemotherapy. Leuk Res 2012; 36(10): 1255-61.

[http://dx.doi.org/10.1016/j.leukres.2012.05.016] [PMID: 22726923]

[19]   Segal R, Zwaal C, Green E, Tomasone JR, Loblaw A, Petrella T. Exercise for people with cancer: a systematic review. Curr Oncol 2017; 24(4): e290-315.
[http://dx.doi.org/10.3747/co.24.3519] [PMID: 28874900]

[20]   Mustian KM, Alfano CM, Heckler C, *et al.* Comparison of Pharmaceutical, Psychological, and Exercise Treatments for Cancer-Related Fatigue: A Meta-analysis. JAMA Oncol 2017; 3(7): 961-8.
[http://dx.doi.org/10.1001/jamaoncol.2016.6914] [PMID: 28253393]

[21]   Rajarajeswaran P, Vishnupriya R. Exercise in cancer. Indian J Med Paediatr Oncol 2009; 30(2): 61-70.
[http://dx.doi.org/10.4103/0971-5851.60050] [PMID: 20596305]

[22]   Sodergren SC, Copson E, White A, *et al.* Systematic Review of the Side Effects Associated With Anti-HER2-Targeted Therapies Used in the Treatment of Breast Cancer, on Behalf of the EORTC Quality of Life Group. Target Oncol 2016; 11(3): 277-92.
[http://dx.doi.org/10.1007/s11523-015-0409-2] [PMID: 26677846]

[23]   Tao JJ, Visvanathan K, Wolff AC. Long term side effects of adjuvant chemotherapy in patients with early breast cancer. Breast 2015; 24 (Suppl. 2): S149-53.
[http://dx.doi.org/10.1016/j.breast.2015.07.035] [PMID: 26299406]

[24]   Ocean AJ, Vahdat LT. Chemotherapy-induced peripheral neuropathy: pathogenesis and emerging therapies. Support Care Cancer 2004; 12(9): 619-25.
[PMID: 15258838]

[25]   Derks MG, de Glas NA, Bastiaannet E, *et al.* Physical Functioning in Older Patients With Breast Cancer: A Prospective Cohort Study in the TEAM Trial. Oncologist 2016; 21(8): 946-53.
[http://dx.doi.org/10.1634/theoncologist.2016-0033] [PMID: 27368882]

[26]   Sehl M, Lu X, Silliman R, Ganz PA. Decline in physical functioning in first 2 years after breast cancer diagnosis predicts 10-year survival in older women. J Cancer Surviv 2013; 7(1): 20-31.
[http://dx.doi.org/10.1007/s11764-012-0239-5] [PMID: 23232922]

[27]   Binkley JM, Harris SR, Levangie PK, *et al.* Patient perspectives on breast cancer treatment side effects and the prospective surveillance model for physical rehabilitation for women with breast cancer. Cancer 2012; 118(8) (Suppl.): 2207-16.
[http://dx.doi.org/10.1002/cncr.27469] [PMID: 22488695]

[28]   Brown JC, Harhay MO, Harhay MN. Physical function as a prognostic biomarker among cancer survivors. Br J Cancer 2015; 112(1): 194-8.
[http://dx.doi.org/10.1038/bjc.2014.568] [PMID: 25393366]

[29]   Merchant CR, Chapman T, Kilbreath SL, Refshauge KM, Krupa K. Decreased muscle strength following management of breast cancer. Disabil Rehabil 2008; 30(15): 1098-105.
[http://dx.doi.org/10.1080/09638280701478512] [PMID: 19230221]

[30]   Klassen O, Schmidt ME, Ulrich CM, *et al.* Muscle strength in breast cancer patients receiving different treatment regimes. J Cachexia Sarcopenia Muscle 2017; 8(2): 305-16.
[http://dx.doi.org/10.1002/jcsm.12165] [PMID: 27896952]

[31]   Neil-Sztramko SE, Kirkham AA, Hung SH, Niksirat N, Nishikawa K, Campbell KL. Aerobic capacity and upper limb strength are reduced in women diagnosed with breast cancer: a systematic review. J Physiother 2014; 60(4): 189-200.
[http://dx.doi.org/10.1016/j.jphys.2014.09.005] [PMID: 25443649]

[32]   Ribeiro Pereira ACP, Koifman RJ, Bergmann A. Incidence and risk factors of lymphedema after breast cancer treatment: 10 years of follow-up. Breast 2017; 36: 67-73.
[http://dx.doi.org/10.1016/j.breast.2017.09.006] [PMID: 28992556]

[33]   Cormie P, Singh B, Hayes S, *et al.* Acute Inflammatory Response to Low-, Moderate-, and High-Load Resistance Exercise in Women With Breast Cancer-Related Lymphedema. Integr Cancer Ther 2016; 15(3): 308-17.

[http://dx.doi.org/10.1177/1534735415617283] [PMID: 26582633]

[34]   Simonavice E, Kim JS, Panton L. Effects of resistance exercise in women with or at risk for breast cancer-related lymphedema. Support Care Cancer 2017; 25(1): 9-15.
[http://dx.doi.org/10.1007/s00520-016-3374-0] [PMID: 27516182]

[35]   Foley MP, Hasson SM. Effects of a Community-Based Multimodal Exercise Program on Health-Related Physical Fitness and Physical Function in Breast Cancer Survivors: A Pilot Study. Integr Cancer Ther 2016; 15(4): 446-54.
[http://dx.doi.org/10.1177/1534735416639716] [PMID: 27151593]

[36]   Chaoul A, Milbury K, Spelman A, *et al.* Randomized trial of Tibetan yoga in patients with breast cancer undergoing chemotherapy. Cancer 2018; 124(1): 36-45.
[http://dx.doi.org/10.1002/cncr.30938] [PMID: 28940301]

[37]   Zou LY, Yang L, He XL, Sun M, Xu JJ. Effects of aerobic exercise on cancer-related fatigue in breast cancer patients receiving chemotherapy: a meta-analysis. Tumour Biol 2014; 35(6): 5659-67.
[http://dx.doi.org/10.1007/s13277-014-1749-8] [PMID: 24570186]

[38]   Mohamady HM, Elsisi HF, Aneis YM. Impact of moderate intensity aerobic exercise on chemotherapy-induced anemia in elderly women with breast cancer: A randomized controlled clinical trial. J Adv Res 2017; 8(1): 7-12.
[http://dx.doi.org/10.1016/j.jare.2016.10.005] [PMID: 27872759]

[39]   Shobeiri F, Masoumi SZ, Nikravesh A, Heidari Moghadam R, Karami M. The Impact of Aerobic Exercise on Quality of Life in Women with Breast Cancer: A Randomized Controlled Trial. J Res Health Sci 2016; 16(3): 127-32.
[PMID: 27840340]

[40]   Evans ES, Battaglini CL, Groff DG, Hackney AC. Aerobic exercise intensity in breast cancer patients: a preliminary investigation. Integr Cancer Ther 2009; 8(2): 139-47.
[http://dx.doi.org/10.1177/1534735409335506] [PMID: 19679622]

[41]   Murtezani A, Ibraimi Z, Bakalli A, Krasniqi S, Disha ED, Kurtishi I. The effect of aerobic exercise on quality of life among breast cancer survivors: a randomized controlled trial. J Cancer Res Ther 2014; 10(3): 658-64.
[PMID: 25313756]

[42]   Hornsby WE, Douglas PS, West MJ, *et al.* Safety and efficacy of aerobic training in operable breast cancer patients receiving neoadjuvant chemotherapy: a phase II randomized trial. Acta Oncol 2014; 53(1): 65-74.
[http://dx.doi.org/10.3109/0284186X.2013.781673] [PMID: 23957716]

[43]   Winters-Stone KM, Dobek J, Bennett JA, Nail LM, Leo MC, Schwartz A. The effect of resistance training on muscle strength and physical function in older, postmenopausal breast cancer survivors: a randomized controlled trial. J Cancer Surviv 2012; 6(2): 189-99.
[http://dx.doi.org/10.1007/s11764-011-0210-x] [PMID: 22193780]

[44]   Winters-Stone KM, Dobek J, Nail L, *et al.* Strength training stops bone loss and builds muscle in postmenopausal breast cancer survivors: a randomized, controlled trial. Breast Cancer Res Treat 2011; 127(2): 447-56.
[http://dx.doi.org/10.1007/s10549-011-1444-z] [PMID: 21424279]

[45]   Simonavice E, Liu PY, Ilich JZ, Kim JS, Arjmandi BH, Panton LB. The Effects of Resistance Training on Physical Function and Quality of Life in Breast Cancer Survivors. Healthcare (Basel) 2015; 3(3): 695-709.
[http://dx.doi.org/10.3390/healthcare3030695] [PMID: 27417791]

[46]   Schmidt ME, Wiskemann J, Armbrust P, Schneeweiss A, Ulrich CM, Steindorf K. Effects of resistance exercise on fatigue and quality of life in breast cancer patients undergoing adjuvant chemotherapy: A randomized controlled trial. Int J Cancer 2015; 137(2): 471-80.
[http://dx.doi.org/10.1002/ijc.29383] [PMID: 25484317]

[47] Cormie P, Galvão DA, Spry N, Newton RU. Neither heavy nor light load resistance exercise acutely exacerbates lymphedema in breast cancer survivor. Integr Cancer Ther 2013; 12(5): 423-32. [http://dx.doi.org/10.1177/1534735413477194] [PMID: 23439658]

[48] American College of Sports Medicine MI. ACSM's Guide to Exercise and Cancer Survivorship 2012.

[49] Courneya KS, Segal RJ, Mackey JR, *et al.* Effects of aerobic and resistance exercise in breast cancer patients receiving adjuvant chemotherapy: a multicenter randomized controlled trial. J Clin Oncol 2007; 25(28): 4396-404. [http://dx.doi.org/10.1200/JCO.2006.08.2024] [PMID: 17785708]

[50] Meneses-Echávez JF, González-Jiménez E, Ramírez-Vélez R. Supervised exercise reduces cancer-related fatigue: a systematic review. J Physiother 2015; 61(1): 3-9. [http://dx.doi.org/10.1016/j.jphys.2014.08.019] [PMID: 25511250]

[51] Fuller JT, Hartland MC, Maloney LT, Davison K. Therapeutic effects of aerobic and resistance exercises for cancer survivors: a systematic review of meta-analyses of clinical trials. Br J Sports Med 2018; 52(20): 1311. [http://dx.doi.org/10.1136/bjsports-2017-098285] [PMID: 29549149]

[52] Dieli-Conwright CM, Courneya KS, Demark-Wahnefried W, *et al.* Effects of Aerobic and Resistance Exercise on Metabolic Syndrome, Sarcopenic Obesity, and Circulating Biomarkers in Overweight or Obese Survivors of Breast Cancer: A Randomized Controlled Trial. J Clin Oncol 2018; 36(9): 875-83. [http://dx.doi.org/10.1200/JCO.2017.75.7526] [PMID: 29356607]

[53] Herrero F, San Juan AF, Fleck SJ, *et al.* Combined aerobic and resistance training in breast cancer survivors: A randomized, controlled pilot trial. Int J Sports Med 2006; 27(7): 573-80. [http://dx.doi.org/10.1055/s-2005-865848] [PMID: 16802254]

[54] Vardar Yağlı N, Şener G, Arıkan H, *et al.* Do yoga and aerobic exercise training have impact on functional capacity, fatigue, peripheral muscle strength, and quality of life in breast cancer survivors? Integr Cancer Ther 2015; 14(2): 125-32. [http://dx.doi.org/10.1177/1534735414565699] [PMID: 25567329]

[55] Lanctôt D, Dupuis G, Marcaurell R, Anestin AS, Bali M. The effects of the Bali Yoga Program (BYP-BC) on reducing psychological symptoms in breast cancer patients receiving chemotherapy: results of a randomized, partially blinded, controlled trial. J Complement Integr Med 2016; 13(4): 405-12. [http://dx.doi.org/10.1515/jcim-2015-0089] [PMID: 27404902]

[56] Şener HÖ, Malkoç M, Ergin G, Karadibak D, Yavuzşen T. Effects of Clinical Pilates Exercises on Patients Developing Lymphedema after Breast Cancer Treatment: A Randomized Clinical Trial. J Breast Health 2017; 13(1): 16-22. [http://dx.doi.org/10.5152/tjbh.2016.3136] [PMID: 28331763]

[57] Espíndula RC, Nadas GB, Rosa MID, Foster C, Araújo FC, Grande AJ. Pilates for breast cancer: A systematic review and meta-analysis. Rev Assoc Med Bras (1992) 2017; 63(11): 1006-12. [http://dx.doi.org/10.1590/1806-9282.63.11.1006] [PMID: 29451666]

[58] Petrylak DP. Chemotherapy for advanced hormone refractory prostate cancer. Urology 1999; 54(6A) (Suppl.): 30-5. [http://dx.doi.org/10.1016/S0090-4295(99)00452-5] [PMID: 10606282]

[59] Saad F, Asselah J. Chemotherapy for prostate cancer: Clinical practice in Canada. Can Urol Assoc J 2013; 7(1-2) (Suppl. 1): S5-S10. [PMID: 23682304]

[60] Watts S, Leydon G, Birch B, *et al.* Depression and anxiety in prostate cancer: a systematic review and meta-analysis of prevalence rates. BMJ Open 2014; 4(3): e003901. [http://dx.doi.org/10.1136/bmjopen-2013-003901] [PMID: 24625637]

[61] Manokumar T, Aziz S, Breunis H, *et al.* A prospective study examining elder-relevant outcomes in older adults with prostate cancer undergoing treatment with chemotherapy or abiraterone. J Geriatr

Oncol 2016; 7(2): 81-9.
[http://dx.doi.org/10.1016/j.jgo.2016.01.003] [PMID: 26853769]

[62]   Levy ME, Perera S, van Londen GJ, Nelson JB, Clay CA, Greenspan SL. Physical function changes in prostate cancer patients on androgen deprivation therapy: a 2-year prospective study. Urology 2008; 71(4): 735-9.
[http://dx.doi.org/10.1016/j.urology.2007.09.018] [PMID: 18279933]

[63]   Clay CA, Perera S, Wagner JM, Miller ME, Nelson JB, Greenspan SL. Physical function in men with prostate cancer on androgen deprivation therapy. Phys Ther 2007; 87(10): 1325-33.
[http://dx.doi.org/10.2522/ptj.20060302] [PMID: 17684084]

[64]   Gonzalez BD, Jim HSL, Small BJ, *et al.* Changes in physical functioning and muscle strength in men receiving androgen deprivation therapy for prostate cancer: a controlled comparison. Support Care Cancer 2016; 24(5): 2201-7.
[http://dx.doi.org/10.1007/s00520-015-3016-y] [PMID: 26563183]

[65]   Segal R. Physical functioning for prostate health. Can Urol Assoc J 2014; 8(7-8) (Suppl. 5): S162-3.
[http://dx.doi.org/10.5489/cuaj.2315] [PMID: 25243044]

[66]   Torti DC, Matheson GO. Exercise and prostate cancer. Sports Med 2004; 34(6): 363-9.
[http://dx.doi.org/10.2165/00007256-200434060-00003] [PMID: 15157121]

[67]   Keogh JW, MacLeod RD. Body composition, physical fitness, functional performance, quality of life, and fatigue benefits of exercise for prostate cancer patients: a systematic review. J Pain Symptom Manage 2012; 43(1): 96-110.
[http://dx.doi.org/10.1016/j.jpainsymman.2011.03.006] [PMID: 21640547]

[68]   Velthuis MJ, Agasi-Idenburg SC, Aufdemkampe G, Wittink HM. The effect of physical exercise on cancer-related fatigue during cancer treatment: a meta-analysis of randomised controlled trials. Clin Oncol (R Coll Radiol) 2010; 22(3): 208-21.
[http://dx.doi.org/10.1016/j.clon.2009.12.005] [PMID: 20110159]

[69]   Dawson JK, Dorff TB, Todd Schroeder E, Lane CJ, Gross ME, Dieli-Conwright CM. Impact of resistance training on body composition and metabolic syndrome variables during androgen deprivation therapy for prostate cancer: a pilot randomized controlled trial. BMC Cancer 2018; 18(1): 368.
[http://dx.doi.org/10.1186/s12885-018-4306-9] [PMID: 29614993]

[70]   Winters-Stone KM, Dobek JC, Bennett JA, *et al.* Resistance training reduces disability in prostate cancer survivors on androgen deprivation therapy: evidence from a randomized controlled trial. Arch Phys Med Rehabil 2015; 96(1): 7-14.
[http://dx.doi.org/10.1016/j.apmr.2014.08.010] [PMID: 25194450]

[71]   Galvão DA, Taaffe DR, Spry N, *et al.* Exercise Preserves Physical Function in Prostate Cancer Patients with Bone Metastases. Med Sci Sports Exerc 2018; 50(3): 393-9.
[http://dx.doi.org/10.1249/MSS.0000000000001454] [PMID: 29036016]

[72]   Yunfeng G, Weiyang H, Xueyang H, Yilong H, Xin G. Exercise overcome adverse effects among prostate cancer patients receiving androgen deprivation therapy: An update meta-analysis. Medicine (Baltimore) 2017; 96(27): e7368.
[http://dx.doi.org/10.1097/MD.0000000000007368] [PMID: 28682886]

[73]   Buffart LM, Newton RU, Chinapaw MJ, *et al.* The effect, moderators, and mediators of resistance and aerobic exercise on health-related quality of life in older long-term survivors of prostate cancer. Cancer 2015; 121(16): 2821-30.
[http://dx.doi.org/10.1002/cncr.29406] [PMID: 25891302]

[74]   Buffart LM, Galvão DA, Chinapaw MJ, *et al.* Mediators of the resistance and aerobic exercise intervention effect on physical and general health in men undergoing androgen deprivation therapy for prostate cancer. Cancer 2014; 120(2): 294-301.
[http://dx.doi.org/10.1002/cncr.28396] [PMID: 24122296]

[75] Molassiotis A, Smith JA, Mazzone P, Blackhall F, Irwin RS, Panel CEC. Symptomatic Treatment of Cough Among Adult Patients With Lung Cancer: CHEST Guideline and Expert Panel Report. Chest 2017; 151(4): 861-74.
[http://dx.doi.org/10.1016/j.chest.2016.12.028] [PMID: 28108179]

[76] Ou SH, Ahn JS, De Petris L, *et al.* Alectinib in Crizotinib-Refractory ALK-Rearranged Non-Small-Cell Lung Cancer: A Phase II Global Study. J Clin Oncol 2016; 34(7): 661-8.
[http://dx.doi.org/10.1200/JCO.2015.63.9443] [PMID: 26598747]

[77] Kazandjian D, Suzman DL, Blumenthal G, *et al.* FDA Approval Summary: Nivolumab for the Treatment of Metastatic Non-Small Cell Lung Cancer With Progression On or After Platinum-Based Chemotherapy. Oncologist 2016; 21(5): 634-42.
[http://dx.doi.org/10.1634/theoncologist.2015-0507] [PMID: 26984449]

[78] Naito T, Okayama T, Aoyama T, *et al.* Skeletal muscle depletion during chemotherapy has a large impact on physical function in elderly Japanese patients with advanced non-small-cell lung cancer. BMC Cancer 2017; 17(1): 571.
[http://dx.doi.org/10.1186/s12885-017-3562-4] [PMID: 28841858]

[79] Kurtz ME, Kurtz JC, Stommel M, Given CW, Given BA. Symptomatology and loss of physical functioning among geriatric patients with lung cancer. J Pain Symptom Manage 2000; 19(4): 249-56.
[http://dx.doi.org/10.1016/S0885-3924(00)00120-2] [PMID: 10799791]

[80] Wintner LM, Giesinger JM, Zabernigg A, *et al.* Quality of life during chemotherapy in lung cancer patients: results across different treatment lines. Br J Cancer 2013; 109(9): 2301-8.
[http://dx.doi.org/10.1038/bjc.2013.585] [PMID: 24091620]

[81] Brown DJ, McMillan DC, Milroy R. The correlation between fatigue, physical function, the systemic inflammatory response, and psychological distress in patients with advanced lung cancer. Cancer 2005; 103(2): 377-82.
[http://dx.doi.org/10.1002/cncr.20777] [PMID: 15558809]

[82] Williams AC, Grant M, Tiep B, Kim JY, Hayter J. Dyspnea Management in Early Stage Lung Cancer: A Palliative Perspective. J Hosp Palliat Nurs 2012; 14(5): 341-2.
[http://dx.doi.org/10.1097/NJH.0b013e318258043a] [PMID: 24058283]

[83] Knols R, Aaronson NK, Uebelhart D, Fransen J, Aufdemkampe G. Physical exercise in cancer patients during and after medical treatment: a systematic review of randomized and controlled clinical trials. J Clin Oncol 2005; 23(16): 3830-42.
[http://dx.doi.org/10.1200/JCO.2005.02.148] [PMID: 15923576]

[84] Granger CL, McDonald CF, Berney S, Chao C, Denehy L. Exercise intervention to improve exercise capacity and health related quality of life for patients with Non-small cell lung cancer: a systematic review. Lung Cancer 2011; 72(2): 139-53.
[http://dx.doi.org/10.1016/j.lungcan.2011.01.006] [PMID: 21316790]

[85] Michaels C. The importance of exercise in lung cancer treatment. Transl Lung Cancer Res 2016; 5(3): 235-8.
[http://dx.doi.org/10.21037/tlcr.2016.03.02] [PMID: 27413700]

[86] Adamsen L, Stage M, Laursen J, Rørth M, Quist M. Exercise and relaxation intervention for patients with advanced lung cancer: a qualitative feasibility study. Scand J Med Sci Sports 2012; 22(6): 804-15.
[http://dx.doi.org/10.1111/j.1600-0838.2011.01323.x] [PMID: 21599754]

[87] Temel JS, Greer JA, Goldberg S, *et al.* A structured exercise program for patients with advanced non-small cell lung cancer. J Thorac Oncol 2009; 4(5): 595-601.
[http://dx.doi.org/10.1097/JTO.0b013e31819d18e5] [PMID: 19276834]

[88] Hwang CL, Yu CJ, Shih JY, Yang PC, Wu YT. Effects of exercise training on exercise capacity in patients with non-small cell lung cancer receiving targeted therapy. Support Care Cancer 2012; 20(12):

3169-77.
[http://dx.doi.org/10.1007/s00520-012-1452-5] [PMID: 22526147]

[89]   Peddle-McIntyre CJ, Bell G, Fenton D, McCargar L, Courneya KS. Feasibility and preliminary efficacy of progressive resistance exercise training in lung cancer survivors. Lung Cancer 2011. [PMID: 21715041]

[90]   Spruit MA, Janssen PP, Willemsen SC, Hochstenbag MM, Wouters EF. Exercise capacity before and after an 8-week multidisciplinary inpatient rehabilitation program in lung cancer patients: a pilot study. Lung Cancer 2006; 52(2): 257-60.
[http://dx.doi.org/10.1016/j.lungcan.2006.01.003] [PMID: 16529844]

[91]   Munker S, Gerken M, Fest P, *et al.* Chemotherapy for metastatic colon cancer: No effect on survival when the dose is reduced due to side effects. BMC Cancer 2018; 18(1): 455.
[http://dx.doi.org/10.1186/s12885-018-4380-z] [PMID: 29685155]

[92]   Ward P, Hecht JR, Wang HJ, *et al.* Physical function and quality of life in frail and/or elderly patients with metastatic colorectal cancer treated with capecitabine and bevacizumab: an exploratory analysis. J Geriatr Oncol 2014; 5(4): 368-75.
[http://dx.doi.org/10.1016/j.jgo.2014.05.002] [PMID: 25002322]

[93]   Brown JC, Zemel BS, Troxel AB, *et al.* Dose-response effects of aerobic exercise on body composition among colon cancer survivors: a randomised controlled trial. Br J Cancer 2017; 117(11): 1614-20.
[http://dx.doi.org/10.1038/bjc.2017.339] [PMID: 28934762]

[94]   Piringer G, Fridrik M, Fridrik A, *et al.* A prospective, multicenter pilot study to investigate the feasibility and safety of a 1-year controlled exercise training after adjuvant chemotherapy in colorectal cancer patients. Support Care Cancer 2018; 26(4): 1345-52.
[http://dx.doi.org/10.1007/s00520-017-3961-8] [PMID: 29168033]

[95]   Morielli AR, Usmani N, Boulé NG, *et al.* Exercise motivation in rectal cancer patients during and after neoadjuvant chemoradiotherapy. Support Care Cancer 2016; 24(7): 2919-26.
[PMID: 26847350]

[96]   Van Vulpen JK, Velthuis MJ, Steins Bisschop CN, *et al.* Effects of an Exercise Program in Colon Cancer Patients undergoing Chemotherapy. Med Sci Sports Exerc 2016; 48(5): 767-75.
[http://dx.doi.org/10.1249/MSS.0000000000000855] [PMID: 26694846]

[97]   van Waart H, Stuiver MM, van Harten WH, *et al.* Recruitment to and pilot results of the PACES randomized trial of physical exercise during adjuvant chemotherapy for colon cancer. Int J Colorectal Dis 2018; 33(1): 29-40.
[http://dx.doi.org/10.1007/s00384-017-2921-6] [PMID: 29124329]

[98]   Courneya KS, Vardy JL, O'Callaghan CJ, *et al.* Effects of a Structured Exercise Program on Physical Activity and Fitness in Colon Cancer Survivors: One Year Feasibility Results from the CHALLENGE Trial. Cancer Epidemiol Biomarkers Prev 2016; 25(6): 969-77.
[http://dx.doi.org/10.1158/1055-9965.EPI-15-1267] [PMID: 27197271]

[99]   Numico G, Longo V, Courthod G, Silvestris N. Cancer survivorship: long-term side-effects of anticancer treatments of gastrointestinal cancer. Curr Opin Oncol 2015; 27(4): 351-7.
[http://dx.doi.org/10.1097/CCO.0000000000000203] [PMID: 26049277]

[100]   Kim HS, Kim JH, Kim JW, Kim BC. Chemotherapy in Elderly Patients with Gastric Cancer. J Cancer 2016; 7(1): 88-94.
[http://dx.doi.org/10.7150/jca.13248] [PMID: 26722364]

[101]   Gilliam LA, St Clair DK. Chemotherapy-induced weakness and fatigue in skeletal muscle: the role of oxidative stress. Antioxid Redox Signal 2011; 15(9): 2543-63.
[http://dx.doi.org/10.1089/ars.2011.3965] [PMID: 21457105]

[102]   Lund M, Alexandersson von Döbeln G, Nilsson M, *et al.* Effects on heart function of neoadjuvant

chemotherapy and chemoradiotherapy in patients with cancer in the esophagus or gastroesophageal junction - a prospective cohort pilot study within a randomized clinical trial. Radiat Oncol 2015; 10: 16.
[http://dx.doi.org/10.1186/s13014-014-0310-7] [PMID: 25582305]

[103] Lalla RV, Sonis ST, Peterson DE. Management of oral mucositis in patients who have cancer. Dent Clin North Am 2008; 52(1): 61-77, viii. [viii.].
[http://dx.doi.org/10.1016/j.cden.2007.10.002] [PMID: 18154865]

[104] O'Neill L, Moran J, Guinan EM, Reynolds JV, Hussey J. Physical decline and its implications in the management of oesophageal and gastric cancer: a systematic review. J Cancer Surviv 2018; 12(4): 601-18.
[http://dx.doi.org/10.1007/s11764-018-0696-6] [PMID: 29796931]

[105] Gockel I, Pfirrmann D, Jansen-Winkeln B, Simon P. The dawning of perioperative care in esophageal cancer. J Thorac Dis 2017; 9(10): 3488-92.
[http://dx.doi.org/10.21037/jtd.2017.09.07] [PMID: 29268323]

[106] Gannon JA, Guinan EM, Doyle SL, Beddy P, Reynolds JV, Hussey J. Reduced fitness and physical functioning are long-term sequelae after curative treatment for esophageal cancer: a matched control study. Dis Esophagus 2017; 30(8): 1-7.
[http://dx.doi.org/10.1093/dote/dox018] [PMID: 28575241]

[107] Breccia M, Alimena G. Occurrence and current management of side effects in chronic myeloid leukemia patients treated frontline with tyrosine kinase inhibitors. Leuk Res 2013; 37(6): 713-20.
[http://dx.doi.org/10.1016/j.leukres.2013.01.021] [PMID: 23473918]

[108] Tong X, Li J, Zhou Z, Zheng D, Liu J, Su C. Efficacy and side-effects of decitabine in treatment of atypical chronic myeloid leukemia. Leuk Lymphoma 2015; 56(6): 1911-3.
[http://dx.doi.org/10.3109/10428194.2014.986477] [PMID: 25426665]

[109] Chang PH, Lai YH, Shun SC, *et al.* Effects of a walking intervention on fatigue-related experiences of hospitalized acute myelogenous leukemia patients undergoing chemotherapy: a randomized controlled trial. J Pain Symptom Manage 2008; 35(5): 524-34.
[http://dx.doi.org/10.1016/j.jpainsymman.2007.06.013] [PMID: 18280104]

[110] Dimeo F, Schwartz S, Fietz T, Wanjura T, Böning D, Thiel E. Effects of endurance training on the physical performance of patients with hematological malignancies during chemotherapy. Support Care Cancer 2003; 11(10): 623-8.
[http://dx.doi.org/10.1007/s00520-003-0512-2] [PMID: 12942360]

[111] Alibhai SM, Breunis H, Timilshina N, *et al.* Quality of life and physical function in adults treated with intensive chemotherapy for acute myeloid leukemia improve over time independent of age. J Geriatr Oncol 2015; 6(4): 262-71.
[http://dx.doi.org/10.1016/j.jgo.2015.04.002] [PMID: 25944029]

[112] Mohamedali H, Breunis H, Timilshina N, *et al.* Older age is associated with similar quality of life and physical function compared to younger age during intensive chemotherapy for acute myeloid leukemia. Leuk Res 2012; 36(10): 1241-8.
[http://dx.doi.org/10.1016/j.leukres.2012.05.020] [PMID: 22727251]

[113] Morishita S, Kaida K, Yamauchi S, *et al.* Gender differences in health-related quality of life, physical function and psychological status among patients in the early phase following allogeneic haematopoietic stem cell transplantation. Psychooncology 2013; 22(5): 1159-66.
[http://dx.doi.org/10.1002/pon.3128] [PMID: 22736382]

[114] Morishita S, Kaida K, Yamauchi S, *et al.* Early-phase differences in health-related quality of life, psychological status, and physical function between human leucocyte antigen-haploidentical and other allogeneic haematopoietic stem cell transplantation recipients. Eur J Oncol Nurs 2015; 19(5): 443-50.
[http://dx.doi.org/10.1016/j.ejon.2015.02.002] [PMID: 25911269]

[115] Jarden M, Adamsen L, Kjeldsen L, *et al.* The emerging role of exercise and health counseling in

patients with acute leukemia undergoing chemotherapy during outpatient management. Leuk Res 2013; 37(2): 155-61.
[http://dx.doi.org/10.1016/j.leukres.2012.09.001] [PMID: 23021021]

[116]   Jarden M, Møller T, Kjeldsen L, *et al.* Patient Activation through Counseling and Exercise--Acute Leukemia (PACE-AL)--a randomized controlled trial. BMC Cancer 2013; 13: 446.
[http://dx.doi.org/10.1186/1471-2407-13-446] [PMID: 24083543]

[117]   Alibhai SM, Durbano S, Breunis H, *et al.* A phase II exercise randomized controlled trial for patients with acute myeloid leukemia undergoing induction chemotherapy. Leuk Res 2015; S0145-2126(15)30365-9.
[PMID: 26350143]

[118]   Oechsle K, Aslan Z, Suesse Y, Jensen W, Bokemeyer C, de Wit M. Multimodal exercise training during myeloablative chemotherapy: a prospective randomized pilot trial. Support Care Cancer 2014; 22(1): 63-9.
[http://dx.doi.org/10.1007/s00520-013-1927-z] [PMID: 23989498]

[119]   Bryant AL, Deal AM, Battaglini CL, *et al.* The Effects of Exercise on Patient-Reported Outcomes and Performance-Based Physical Function in Adults With Acute Leukemia Undergoing Induction Therapy: Exercise and Quality of Life in Acute Leukemia (EQUAL). Integr Cancer Ther 2018; 17(2): 263-70.
[http://dx.doi.org/10.1177/1534735417699881] [PMID: 28627275]

[120]   Battaglini CL, Hackney AC, Garcia R, Groff D, Evans E, Shea T. The effects of an exercise program in leukemia patients. Integr Cancer Ther 2009; 8(2): 130-8.
[http://dx.doi.org/10.1177/1534735409334266] [PMID: 19679621]

[121]   Sitaresmi MN, Mostert S, Purwanto I, Gundy CM, Sutaryo , Veerman AJ. Chemotherapy-related side effects in childhood acute lymphoblastic leukemia in Indonesia: parental perceptions. J Pediatr Oncol Nurs 2009; 26(4): 198-207.
[http://dx.doi.org/10.1177/1043454209340315] [PMID: 19726791]

[122]   Ramphal R, Aubin S, Czaykowski P, *et al.* Adolescent and young adult cancer: principles of care. Curr Oncol 2016; 23(3): 204-9.
[http://dx.doi.org/10.3747/co.23.3013] [PMID: 27330350]

[123]   Ness KK, Armenian SH, Kadan-Lottick N, Gurney JG. Adverse effects of treatment in childhood acute lymphoblastic leukemia: general overview and implications for long-term cardiac health. Expert Rev Hematol 2011; 4(2): 185-97.
[http://dx.doi.org/10.1586/ehm.11.8] [PMID: 21495928]

[124]   Langeveld N, Ubbink M, Smets E, Group DLES. 'I don't have any energy': The experience of fatigue in young adult survivors of childhood cancer. Eur J Oncol Nurs 2000; 4(1): 20-8.
[http://dx.doi.org/10.1054/ejon.1999.0063] [PMID: 12849627]

[125]   Bryant R. Managing side effects of childhood cancer treatment. J Pediatr Nurs 2003; 18(2): 113-25.
[http://dx.doi.org/10.1053/jpdn.2003.11] [PMID: 12720208]

[126]   van Brussel M, Takken T, van der Net J, *et al.* Physical function and fitness in long-term survivors of childhood leukaemia. Pediatr Rehabil 2006; 9(3): 267-74.
[http://dx.doi.org/10.1080/13638490500523150] [PMID: 17050404]

[127]   Christiansen JR, Kanellopoulos A, Lund MB, *et al.* Impaired exercise capacity and left ventricular function in long-term adult survivors of childhood acute lymphoblastic leukemia. Pediatr Blood Cancer 2015; 62(8): 1437-43.
[http://dx.doi.org/10.1002/pbc.25492] [PMID: 25832752]

[128]   Baumann FT, Bloch W, Beulertz J. Clinical exercise interventions in pediatric oncology: a systematic review. Pediatr Res 2013; 74(4): 366-74.
[http://dx.doi.org/10.1038/pr.2013.123] [PMID: 23857296]

[129] Braam KI, van der Torre P, Takken T, Veening MA, van Dulmen-den Broeder E, Kaspers GJ. Physical exercise training interventions for children and young adults during and after treatment for childhood cancer. Cochrane Database Syst Rev 2016; 3: CD008796. [PMID: 27030386]

[130] Yeh CH, Man Wai JP, Lin US, Chiang YC. A pilot study to examine the feasibility and effects of a home-based aerobic program on reducing fatigue in children with acute lymphoblastic leukemia. Cancer Nurs 2011; 34(1): 3-12. [http://dx.doi.org/10.1097/NCC.0b013e3181e4553c] [PMID: 20706112]

[131] Marchese VG, Chiarello LA, Lange BJ. Effects of physical therapy intervention for children with acute lymphoblastic leukemia. Pediatr Blood Cancer 2004; 42(2): 127-33. [http://dx.doi.org/10.1002/pbc.10481] [PMID: 14752875]

[132] Andrews S, von Gruenigen VE. Management of the late effects of treatments for gynecological cancer. Curr Opin Oncol 2013; 25(5): 566-70. [http://dx.doi.org/10.1097/CCO.0b013e328363e11a] [PMID: 23942302]

[133] Lundqvist EÅ, Fujiwara K, Seoud M. Principles of chemotherapy. Int J Gynaecol Obstet 2015; 131 (Suppl. 2): S146-9. [http://dx.doi.org/10.1016/j.ijgo.2015.06.011] [PMID: 26433671]

[134] Wenzel L, Vergote I, Cella D. Quality of life in patients receiving treatment for gynecologic malignancies: special considerations for patient care. Int J Gynaecol Obstet 2003; 83 (Suppl. 1): 211-29. [http://dx.doi.org/10.1016/S0020-7292(03)90123-8] [PMID: 14763177]

[135] Rutledge TL, Rogers R, Lee SJ, Muller CY. A pilot randomized control trial to evaluate pelvic floor muscle training for urinary incontinence among gynecologic cancer survivors. Gynecol Oncol 2014; 132(1): 154-8. [http://dx.doi.org/10.1016/j.ygyno.2013.10.024] [PMID: 24183730]

[136] Newton MJ, Hayes SC, Janda M, *et al.* Safety, feasibility and effects of an individualised walking intervention for women undergoing chemotherapy for ovarian cancer: a pilot study. BMC Cancer 2011; 11: 389. [http://dx.doi.org/10.1186/1471-2407-11-389] [PMID: 21899778]

[137] Zhang Q, Li F, Zhang H, Yu X, Cong Y. Effects of nurse-led home-based exercise & cognitive behavioral therapy on reducing cancer-related fatigue in patients with ovarian cancer during and after chemotherapy: A randomized controlled trial. Int J Nurs Stud 2018; 78: 52-60. [http://dx.doi.org/10.1016/j.ijnurstu.2017.08.010] [PMID: 28939343]

[138] McDermott MM, Ades P, Guralnik JM, *et al.* Treadmill exercise and resistance training in patients with peripheral arterial disease with and without intermittent claudication: a randomized controlled trial. JAMA 2009; 301(2): 165-74. [http://dx.doi.org/10.1001/jama.2008.962] [PMID: 19141764]

[139] Daly LE, Ní Bhuachalla ÉB, Power DG, Cushen SJ, James K, Ryan AM. Loss of skeletal muscle during systemic chemotherapy is prognostic of poor survival in patients with foregut cancer. J Cachexia Sarcopenia Muscle 2018; 9(2): 315-25. [http://dx.doi.org/10.1002/jcsm.12267] [PMID: 29318756]

[140] Van Moll CC, Schep G, Vreugdenhil A, Savelberg HH, Husson O. The effect of training during treatment with chemotherapy on muscle strength and endurance capacity: A systematic review. Acta Oncol 2016; 55(5): 539-46. [http://dx.doi.org/10.3109/0284186X.2015.1127414] [PMID: 26755191]

[141] Elter T, Stipanov M, Heuser E, *et al.* Is physical exercise possible in patients with critical cytopenia undergoing intensive chemotherapy for acute leukaemia or aggressive lymphoma? Int J Hematol 2009; 90(2): 199-204. [http://dx.doi.org/10.1007/s12185-009-0376-4] [PMID: 19629631]

[142] Mina DS, Langelier D, Adams SC, *et al.* Exercise as part of routine cancer care. Lancet Oncol 2018; 19(9): e433-6.
[http://dx.doi.org/10.1016/S1470-2045(18)30599-0] [PMID: 30191843]

[143] Fu JB, Tennison JM, Rutzen-Lopez IM, *et al.* Bleeding frequency and characteristics among hematologic malignancy inpatient rehabilitation patients with severe thrombocytopenia. Support Care Cancer 2018; 26(9): 3135-41.
[http://dx.doi.org/10.1007/s00520-018-4160-y] [PMID: 29594490]

[144] Ibanez K, Espiritu N, Souverain RL, *et al.* Safety and Feasibility of Rehabilitation Interventions in Children Undergoing Hematopoietic Stem Cell Transplant With Thrombocytopenia. Arch Phys Med Rehabil 2018; 99(2): 226-33.
[http://dx.doi.org/10.1016/j.apmr.2017.06.034] [PMID: 28807693]

[145] Morishita S, Kaida K, Setogawa K, *et al.* Safety and feasibility of physical therapy in cytopenic patients during allogeneic haematopoietic stem cell transplantation. Eur J Cancer Care (Engl) 2013; 22(3): 289-99.
[http://dx.doi.org/10.1111/ecc.12027] [PMID: 23252444]

[146] Schmitz KH, Courneya KS, Matthews C, *et al.* American College of Sports Medicine roundtable on exercise guidelines for cancer survivors. Med Sci Sports Exerc 2010; 42(7): 1409-26.
[http://dx.doi.org/10.1249/MSS.0b013e3181e0c112] [PMID: 20559064]

[147] Courneya KS. Exercise guidelines for cancer survivors: are fitness and quality-of-life benefits enough to change practice? Curr Oncol 2017; 24(1): 8-9.
[http://dx.doi.org/10.3747/co.24.3545] [PMID: 28270718]

[148] Wolin KY, Schwartz AL, Matthews CE, Courneya KS, Schmitz KH. Implementing the exercise guidelines for cancer survivors. J Support Oncol 2012; 10(5): 171-7.
[http://dx.doi.org/10.1016/j.suponc.2012.02.001] [PMID: 22579268]

[149] Morishita S, Tsubaki A, Fu JB. Does physical activity improve survival and mortality among patients with different types of cancer? Future Oncol 2017; 13(12): 1053-5.
[http://dx.doi.org/10.2217/fon-2017-0037] [PMID: 28492089]

[150] Scott JM, Li N, Liu Q, *et al.* Association of Exercise With Mortality in Adult Survivors of Childhood Cancer. JAMA Oncol 2018; 4(10): 1352-8.
[http://dx.doi.org/10.1001/jamaoncol.2018.2254] [PMID: 29862412]

[151] Hayes SC, Steele ML, Spence RR, *et al.* Exercise following breast cancer: exploratory survival analyses of two randomised, controlled trials. Breast Cancer Res Treat 2018; 167(2): 505-14.
[http://dx.doi.org/10.1007/s10549-017-4541-9] [PMID: 29063309]

[152] Ammitzbøll G, Søgaard K, Karlsen RV, *et al.* Physical activity and survival in breast cancer. Eur J Cancer 2016; 66: 67-74.
[http://dx.doi.org/10.1016/j.ejca.2016.07.010] [PMID: 27529756]

[153] Meyerhardt JA, Giovannucci EL, Holmes MD, *et al.* Physical activity and survival after colorectal cancer diagnosis. J Clin Oncol 2006; 24(22): 3527-34.
[http://dx.doi.org/10.1200/JCO.2006.06.0855] [PMID: 16822844]

CHAPTER 3

Cancer Immune Evasion in Gastrointestinal Cancer: Can this be Overcome by Combination of Histone Deacetylase and Immune Checkpoint Inhibitors?

Yuequan Shi[1,$*], **Onyinyechi Duru[2,$]**, **Zifang Zou[1]** and **David Kerr[3,#*]**

[1] *China Medical University, Liaoning, China*

[2] *Department of Oncology, Nottingham University Hospital, City Campus, Nottingham, UK*

[3] *Radcliffe Department of Medicine, University of Oxford, UK*

Abstract: Although immune checkpoint inhibitors have been used for the treatment of gastrointestinal malignancies, clinical benefit has been modest compared to other tumour sites. Microsatellites high cancers have shown a better response to immunotherapy but they make up a small percentage of this group of tumours. Efforts to improve results have been made, particularly through the use of combination therapy and such an example would be the combination with anti-VEGF agents which have shown some promise in gastric cancers. Histone deacetylase inhibitors (HDACi) have been shown to be effective anti-cancer agents particularly for haematological malignancies with multiple anti-tumour mechanisms of action including the induction of apoptosis, cell cycle arrest and the upregulation of major histocompatibility complex (MHC) in tumour cells. Although there has not been much clinical success in the context of gastrointestinal cancers, there is preclinical evidence to suggest that combination therapy with traditional chemotherapy agents may have some therapeutic benefit. However, the combination of HDACi with immune checkpoint inhibitors has not been studied. Both HDACi and immune checkpoint inhibitors have already demonstrated satisfactory safety profiles and furthermore, clinical activity of HDACi can be monitored by the use of biomarkers. Therefore, it has been hypothesised that by combining the two treatment agents together, synergism may be observed in the form of improved host immune anti- tumour response as a result of enhanced immunogenicity conferred by HDACi, which will ultimately result in effective tumour killing.

* **Corresponding author Yuequan Shi:** China Medical University, Liaoning, China; E-mail: yuequanshi1202@gmail.com
* **Corresponding author David Kerr:** Radcliffe Department of Medicine, University of Oxford, UK; E-mail: david.kerr@ndcls.ox.ac.uk
# To whom requests for reprints be addressed
$ Both authors contributed equally to this paper

Keywords: Cancer, Histone Deacetylase, Immune Checkpoint Blockade, PD1.

INTRODUCTION

Immunotherapy, working through immune checkpoint blockade has achieved notable responses in multiple tumor types including malignant melanoma, renal cell carcinoma, non-small cell lung cancer, bladder carcinoma, Hodgkin's lymphoma, triple-negative breast carcinoma as well as head and neck cancer. However, colorectal cancer (CRC) appears to be one of the tumor types that shows a poor response to immune checkpoint inhibitors, apart from the Microsatellite Unstable (MSI-High) CRC subtype, which only accounts for about 5% of advanced and metastatic CRC.

So why don't immune checkpoint inhibitors work in all tumour types? In this brief review we will consider the following:

a. Mechanisms underlying immune escape by tumours through immunoediting.
b. Focus on clinical trials of gastrointestinal and pancreatic cancer with immune checkpoint inhibitors.
c. Hypothesize that histone deacetylase (HDAC) inhibitors reverse immunoediting to some extent and can therefore function as potentially synergistic combination partners for immune-oncology agents.

MECHANISMS OF TUMOUR IMMUNE EVASION

Successful tumour growth can be attributed to the well-recognized hallmarks of cancer which include sustained proliferation; evasion of growth suppressors; resistance of apoptosis; immortality; invasion and metastases; angiogenesis; alteration of normal cellular metabolism and avoiding detection and destruction by the immune system [1]. To understand the mechanism by which tumours are able to evade the immune system, one must first be cognizant of the processes involved in mounting a successful immune response.

Immunogenicity describes the ability of a substance to be able to provoke an immune response within a host which is executed by either the innate or adaptive immune system and central to both is the ability to recognize and destroy pathogens. Briefly, the innate system, being relatively simpler, recognizes generic peptides common to pathogens and mounts a response, mediated by cytokines and granulocytes within moments. The adaptive immune system is more specific and is activated by processed antigen peptides loaded onto major histocompatibility complex (MHC) I or II molecules found on the surfaces of all cells or specific antigen presenting cells (APCs), which are then presented to either B or T lymphocytes to produce a humoral or cellular response, respectively. The result is a more targeted and robust response and subsequently, the development of memory for that particular pathogen [2]. The cellular arm of the adaptive response is a highly regulated multi-step process beginning with assembly and peptide loading of the MHC complex and its subsequent transportation to the cell surface to present to T-cells. Co-receptors on the T-cell surface, either stimulatory (CD28) or inhibitory (CTLA-4 or PD-1) in nature, interact with their respective ligands on the APC surface resulting in a balanced immune response [3]. Furthermore, the intensity of a response can be modified by a subset of CD4 T-cells expressing CD25 and FoxP3, termed regulatory T-cells, which act to appropriately diminish an effector immune response if necessary [4]. Fig. (1) illustrates the many receptors on both the T-cell and APC surface which can interact and ultimately lead to a controlled immune response [5].

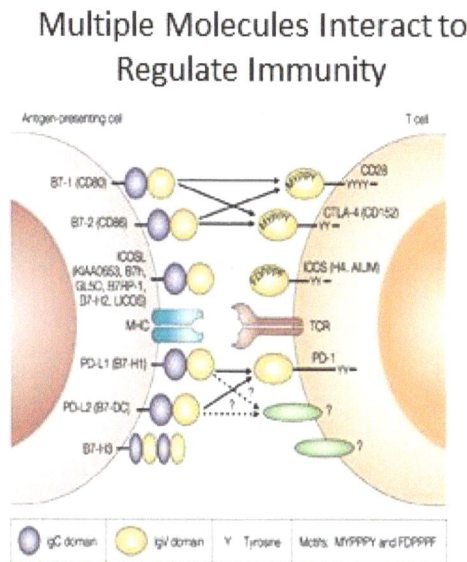

Fig. (1). Cell surface receptors on T-cell and antigen presenting cells which are involved in mounting a controlled immune response.

Immunoediting describes the dual role of the immune system providing host protection from tumours and subsequently sculpting tumours as a result of reduced immunogenicity. This process is comprised of three steps: elimination; equilibrium and escape. In the elimination phase, both the innate and adaptive immune system work in synergy to detect and destroy tumour cells [6, 7]. However, some cells survive and enter the equilibrium phase, whereby the immune system exerts immunologic pressure to prevent further growth of the surviving population [8, 9]. Those tumour cells that are able to resist this pressure, due to genetic or epigenetic changes in their DNA, enter the escape phase to become well established and clinically apparent cancers. Multiple mechanisms to facilitate immunoediting have been proposed and these can be divided into direct cellular interactions between tumour cells and immune cells or manipulation of the surrounding tumour environment.

Direct Cellular Interactions

Altered Expression of Co-receptor Molecules

T-cell activation is a two-step process which relies on signaling from the interaction between T-cell receptors and peptide loaded MHC complexes on the APC, followed by signals from the interaction between the CD28 co-stimulating receptor and its ligands CD80 or CD86. Both CD80 and CD86 belong to the B7 family of molecules, an umbrella term used to describe the ligands that interact with either the co-stimulatory or co-inhibitory receptors on the T-cell surface. Without the second co-stimulatory signal, T-cells are not only unable to become activated but are also rendered anergic [10]. Down regulation of co-stimulatory molecules is a strategy employed by tumor cells to reduce their immunogenicity and indeed decreased expression of co-stimulatory molecules has been observed in different tumours [11 - 15]. Conversely, up-regulation of co-inhibitory molecules can have the same effect. PD-1, a co-inhibitory receptor on the T-cell surface plays a role in modifying T-cell responses when engaged with its ligand PD-L1/B7-H1. Studies have shown that PD-1 is over-expressed on cytotoxic CD8 cells that are no longer functional as a result of exhaustion from chronic viral stimulation [16, 17]. Furthermore, PD-1 knocks out mice are susceptible to autoimmune disease, implicating PD-1 receptor activation in the negative control of immune responses [18]. Another member of the B7 family, B7-H4, suppresses immune activity by reducing cell proliferation and cytokine production of T-cells [19]. Over expression of B7-H1 and B7-H4 have been described in numerous tumours including esophageal [20], gastric [21, 22] and pancreatic cancer [23] and more recently, it has been shown that B7-H4 is associated with poor prognosis in patients with pancreatic cancer [24]

Fas/Fas-ligand Interactions

Fas receptor is expressed on immune and non -immune cells and once activated by its ligand Fas-L, initiates apoptosis within the expressing cell. It is possible for a cell to possess both the receptor and ligand at the same time rendering itself a potential victim or instigator of Fas induced apoptosis. Cytotoxic T cells bear both Fas-L and Fas receptor on their surface however, activated T cells up regulate Fas-L and along with the release of cytotoxic perforin, use the Fas-L/Fas mode of killing to execute their effector function. Tumours are able provide a counter attack and protect themselves by over-expressing Fas-L which induces apoptosis in the Fas receptor bearing lymphocytes. It has previously been shown that upregulation of Fas-L in colon adenocarcinomas can kill lymphocytes and is a means of immune evasion by cancer cells [25].

Interference with MHC Complex Expression and Assembly

As has already been mentioned, presentation of antigen peptides on MHC molecules, class I or II presented to either CD8 or CD4 T-cells respectively is the first required signal for T-cell activation and subsequent cytotoxic response. Human-Leucocyte Antigen cells (HLA) are the human equivalent and both terms are used interchangeably. Successful presentation to the T lymphocytes is dependent on the structural integrity of the HLA complex, loading and transportation to the cell surface and proteins such as β_{2-} macroglobulin, transporter associated with antigen processing (TAP) and low molecular weight protein (LMP) are implicated. In the event that HLA-class I molecules are absent, down-regulated or aberrant, natural killer cells (NK), which are lymphocytes from the innate immune system, become activated as a consequence of loss of recognition of 'self' HLA molecules which would otherwise inhibit NK activation [26] highlighting the interplay between the innate and adaptive immune system. The implications of this in the context of immunoediting are as follows: 1) The down –regulation of HLA complexes initially results in the destruction of tumour cells by NK cells in the elimination phase; 2) Surviving tumour cells maintain a reduced number of HLA molecules to prevent efficient HLA presentation of antigens to T-cells whilst simultaneously keeping this reduced expression above a particular threshold, enabling them to escape recognition by natural killer cells.

Down-regulation of HLA complexes has been reported in colorectal cancer but the mechanism through which this occurs is thought to differ between microsatellite positive (MSI +ve) and microsatellite negative (MSI-ve)/ microsatellite stable (MSS) tumours and this will be explained later. Firstly, microsatellite instability (MSI) is a consequence of DNA mismatch repair deficiency (MMR) and is characterized by increased frequency of deletion or insertion mutations in simple repeat sequences and MSI is the driving force behind tumorigenesis of some colorectal cancers. The high mutation rate may increase the tumour immunogenicity as the tumour is more likely to produce neo-antigens which will be easily recognized by the immune system and mount an immune response and indeed it has been shown that MSI+ve tumours are associated with strong lymphocyte infiltration [27] and patients have a much better prognosis compared to their microsatellite stable (MSS) counterparts [28]. In relation to the HLA down regulation, a study has shown a correlation between

β_2 macroglobulin mutation and total HLA class I loss in MSI positive colorectal tumour cells [29]. Interestingly however, the authors of that study found that the total loss of HLA I expression in MSS tumours was associated with downregulation of LMP7 and TAP2.

There are data suggesting that downregulation of MHC class I expression correlates with a poor prognosis. Watson and colleagues measured MHC I expression in a series of 450 colorectal cancers on a tissue microarray and demonstrated that tumours with low-level expression had a significantly worse prognosis on multivariate modelling. The authors speculated that these tumours would avoid both NK and T cell mediated immune surveillance, and that they would be unlikely to respond to immunotherapy [30].

Tumour Environment Manipulation

Tumour Microenvironment

Normal physiology dictates that breached barriers within a host would result in an inflammatory response orchestrated by cells from the innate immune system such as macrophages or dendritic cells which either release pro-inflammatory cytokines or become mature and engage the adaptive immune system respectively. Naturally, therefore, one would expect the presence of leukocytes in tissue that has been compromised by cancerous cells. However, the effector function of leukocytes such as macrophages is fluid and can alternate between a pro- or anti-inflammatory phenotype, which is determined by surrounding cytokines. Tumour associated macrophages (TAMs) exist as two subtypes; M1 which produce pro-inflammatory cytokines; M2 which are anti-inflammatory and release cytokines that promote tumour growth such as TGF- β and IL-10. TAMs can be polarized to the M2 phenotype under the influence of lactic acid, which is a byproduct produced by tumour cells utilizing anaerobic glycolysis in hypoxic conditions [31] and a high preponderance of M2 TAMs is associated with poor prognosis in gastric cancer [32].

Regulatory T-Cell Invasion

CD4 T-cells expressing FoxP3 and CD25, termed regulatory T-cells, dampen down immune responses when activated and numerous studies have shown a correlation between high T-reg infiltration rates and poor prognosis in many tumour types including colorectal cancer [33 - 36]. However, the impact of T-reg cells on tumorigenesis has not been clearly elucidated as the current literature suggests opposing roles of this subset of cells. One study has shown that tumours from stage II and stage III colorectal cancer with high T-reg infiltration in normal mucosa were associated with poor prognosis compared to high T-reg density within tumour tissue where improved survival was observed [37]. Following this observation, the authors of a review postulated that this difference could be due to the influence of the tumour environment on T-reg cells depending on localization in and around the tumour [38]. Nonetheless, the overall view is that an interaction exists between the tumour environment and T-reg cells, and it has been proposed that in murine models with colorectal cancer, recruitment of T-reg cells is as a result of chemokines secreted by TAMs [39], which upon arrival exert their immunosuppressive action through the secretion of TGF- β. Unsurprisingly perhaps, large numbers of T-reg cells have been associated with poor prognosis in gastric cancer [40] and pancreatic cancer [41].

IMMUNE CHECKPOINT INHIBITORS

Mechanisms of Action of Immune Checkpoint Inhibitors

As mentioned before, down regulation of co-stimulatory molecules and up-regulation of co-inhibitory molecules have the potential to reduce tumour immunogenicity. Currently, cytotoxic T lymphocyte antigen protein-4 (CTLA-4), programmed death receptor-1 (PD-1) and programmed death receptor ligand-1 (PD-L1) are the 3 major types of targeted molecules that have undergone significant clinical evaluation. Unlike other anti-cancer therapies that focus directly on cancer cells, immune checkpoint inhibitors work on blocking the interaction between lymphocytes and the antigen presenting cells (APC) or tumor cells.

CTLA-4 and Anti CTLA-4 Therapy

CTLA-4, also called CD152, is the first identified co-inhibitory receptor exclusively expressed on activated T cell that shares homologous structure with CD28 but has a much higher affinity in binding with both of the ligands CD80 (also known as B7.1) or CD86 (also known as B7.2) on APC [42, 43] and functions predominantly in the T cell priming stage in lymph nodes.

It has been proposed that once a tumor cell antigen is recognized by immune system, both the antigen-specific signals and costimulatory signals (e.g. CD28 or inducible costimulatory signal (ICOS)) transmitted by co-receptors are needed in T-cell activation. According to the cancer immunity cycle framework constructed by Chen and Mellman, this comprises of 7 steps: the release of neoantigens into the tumour microenvironment: presentation of neoantigens to antigen presenting cells (APC): the priming and activation of effector T cell responses: sequestration of T cells to tumours: infiltration of the tumour bed by T cells traversing the through blood vessels: recognition of tumour cells: killing of the malignant cells [44].

With higher affinity than CD28, CTLA-4 plays a negative role in T cell responses which consequently causes tumor progression [44].

Moreover, independently from modulating the activity of effector T cells, it has been proved that CTLA-4 plays a role in enhancing immune suppressive activity of regulatory T cells which constitutively express CTLA-4 [45]. Also, CTLA-4 has a physiological role in down modulating the activity of CD4+ helper T cells which may also contribute to the underlying mechanism of anti-CTLA-4 therapy [46, 47].

Two anti-CTLA-4 monoclonal antibodies, Ipilimumab and Tremelimumab, are undergoing multiple clinical trials in a variety of cancer types and Ipilimumab was the first immune checkpoint inhibitor approved by US food and drug administration in the treatment of metastatic melanoma [48].

PD-1/PD-L1 and Anti-PD-1/PD-L1 Therapy

PD-1, also called CD279, is a co-inhibitory receptor expressed on cell surface of CD4+ and CD8+ T lymphocytes, B lymphocytes, natural killer cells (NK cells), monocytes, dendritic cells (DC) and tumor infiltrating lymphocytes (TIL) [49]. It has 2 ligands: PD-L1 (B7-H1/CD274), expressed both on APC and solid tumor cells, and PD-L2, expressed only on APC cells. Under normal circumstances, interaction between PD-1 and its ligands can lead to inactivation of T cell response thus avoiding an over inflammatory response in peripheral tissue and suppressing autoimmunity [50, 51].

Unlike CTLA-4, PD1/PD-L1 affect T cell interactions with APC in lymph nodes in addition to T cell interactions with cancer cells in the tumor microenvironment [44]. It has been reported that PD-1 is largely expressed on tumor infiltrating lymphocytes in the tumor microenvironment in pancreatic cancer resulting in the T-cell apoptosis which permits tumor growth [52, 53].

Anti-PD-1(e.g. Nivolumab and Pembrolizumab, human monoclonal antibodies that blocks PD-1 receptor) and anti-PD-L1 (e.g. Atezolizumab, a monoclonal human anti-PD-L1 antibody) therapy are the most robust immunotherapies being tested currently in multiple types of cancer and were approved by FDA in the treatment of melanoma, non-small cell lung cancer, Hodgkin lymphoma, Microstallite Unstable cancers (MSI-High)etc. Anti-PD-1 therapy aims to block binding between PD-1 and its ligand PD-L1, therefore blocking the negative signaling and restoring T cell function and therefore anticancer properties. Apart from binding to PD-1, PD-L1 can also interact with B7-1 molecule which transmit negative signal to T cell immunity. Therefore, anti-PD-L1 therapy may have more effects than targeting PD-1 [54].

Clinical Trials of Immune Checkpoint Inhibitors in Gastrointestinal and Pancreatic Tumours

Gastric Cancer

Dating back to 2010, a phase II trial aimed at investigating the safety, clinical efficacy, and immunologic activity of the anti-CTLA4 antibody Tremelimumab was designed and initiated in metastatic gastric and esophageal adenocarcinoma. Of all the 18 patients who received Tremelimumab every 3 months, only 1 patient achieved a significantly durable clinical benefit [55].

Gastric cancer has the highest preponderance in the Japanese population globally. Avelumab (MSB0010718C) is a fully human anti-PD-L1 IgG1 antibody. A Phase Ib dose expansion study was conducted to assess the safety and clinical efficacy of Avelumab in Japanese advanced gastric cancer patients. Patients received Avelumab at 10mg/kg every 2 weeks. Safety can be protected by such dose. Of all the 20 patients who received 6 months treatment, confirmed objective response rate (ORR) was 15% and the disease control rate (DCR; patients with responding and stable disease) was 65%. PD-L1 expression was evaluable in 19 patients. Higher ORR and progression-free survival (PFS) rate were observed in PD-L1+ve patients in contrast with the (-ve) group, suggesting that patient selection based on PD-L1 expression could be of clinical benefit [56, 57].

Subsequent follow up of a larger patient cohort(n= 151 patients 62 of whom were treated with Avelumab as the second-line therapy (2L) while 89 patients took it as first-line maintenance therapy (Mn)). 15 patients reported Grade>=3 treatment related adverse events, and 1 developed treatment-related death (TRD) (hepatic failure/ autoimmune hepatitis). DCR was 29.0% in 2L group and 57.3% in Mn group. Median PFS was 6.0 weeks and 12 weeks respectively. PD-L1 expression could be assessed in 74 patients in total. The data above led to the conclusion that single-agent Avelumab was safe and offered a promising treatment for gastric or gastro-esophageal junction (GEJ) cancer. Two randomized phase III trials are still ongoing [58].

The multicenter, open-label, phase 1b trial KEYNOTE-012 aimed to assess the safety and efficacy of the anti-PD-1 antibody Pembrolizumab. 39 Patients with PD-1 positive tumours who suffered from recurrent or metastatic adenocarcinoma of the stomach or GEJ were treated with Pembrolizumab at 10mg/kg once every 2 weeks. 8 out of 36 patients with evaluable disease showed an overall response. 5 out of 39 patients developed grade 3 or 4 treatment-related adverse events (TRAE) suggesting that this agent is worth pursuing in randomized trials [59].

Charles *et al* designed the large multicohort phase II trial KEYNOTE-059 to assess the efficacy of Pembrolizumab monotherapy in contrast with the Pembrolizumab+5-fluorouracil (5-FU) and cisplatin. 259 Patients received Pembrolizumab 200mg every 3 weeks. The primary endpoints were ORR, safety and tolerability. The safety data were previously published in June 2016, showing that 25 patients experienced drug-related adverse events (DRAE), of which70% were grade 3-4. However, no patients discontinued treatment due to DRAEs, which indicates a manageable side effect profile of Pembrolizumab [60].

At data cutoff in Oct 2016, of all the 259 patients, 55% (n=143) expressed PD-L1. The PD-L1 staining was evaluated by the PD-L1 IHC 22C3 pharmDx Kit (Dako), and the results analyzed in those two groups were summarized in Table **1** (Partial Results of KEYNOTE-059). The total overall ORR was 11.2%, 1.9% of the patients had complete response (CR), 9.3% had partial response (PR), 17% had stable disease (SD) and 55.6% had progressive disease (PD) [61].

Table 1. Partial results of KEYNOTE-059.

	ORR	CR	PR
PD-L1+ pts	15.5%	2%	13.5%
PD-L1- pts	5.5%	1.8%	3.7%

Finally, on September 22, 2017, the Food and Drug Administration granted accelerated approval to Pembrolizumab for patients with recurrent locally advanced or metastatic, gastric or GEJ adenocarcinoma whose tumor express PD-L1 as determined by an FDA-approved test. And patients must have had disease progression on or after two or more prior systemic therapies.

Being aware of the promising antitumor activity of Pembrolizumab as a monotherapy in gastric cancer, a phase 1b/2, open label, dose-expansion study of Margetuximab (M) an Fc-optimized monoclonal antibody that targets the human epidermal growth factor receptor 2, in combination with Pembrolizumab (P) in patients with relapsed/refractory advanced HER2+ GEJ or gastric cancer were initiated in January 2016. M is administered at 2 dose levels: 10mg/kg and 15mg/kg, with a fixed dose of P (200mg). The study is still ongoing in North America and Asia [62].

There are data to suggest that inhibition of angiogenesis can offer useful disease palliation for gastric cancer patients. This suggested that the anti-Vascular Endothelial Growth Factor Inhibitor, Ramucirumab would be a logical combination partner with Pembrolizumab (P). Chau I *et al* designed a multi-cohort, phase 1a/b trial which enrolled patients with pretreated gastric/ GEJ adenocarcinoma. Cohort A were treated with Ramucirumab(R) 8mg/kg on Days

1&8. Cohort B received R 10mg/kg on Day1, and both cohorts were given P 200mg on Day 1 every 3 weeks. The safety of adding R to P was considered fairly well tolerated and preliminary analysis showed 3 of 40 (7.5%) pts (PD-L1 negative, n=1; PD-L1 positive, n=2) have responded to treatment with a 45% disease control rate. Median PFS was 2.10 months (95% CI, 1.18 to 4.04) and 2.60 months (1.38, NR) for Cohorts A and B, respectively [63].

In February 2016, and May 2016, results from the phase I/II open-label CHECKMATE-032 study were published. Part of the study to investigate the efficacy of Nivolumab monotherapy in patients with solid tumors include advanced gastric or GEJ cancer. 59 patients (more than 80% of the patients were pretreated) were given Nivolumab intravenously (3mg/Kg every 2 weeks). As a result, ORR was 12%, including 1 patient who achieved CR and 6 patients with PR. Median overall survival was about 7 months and the 12 months' overall survival rate was 38%. Grade 3 TRAE developed only in 14% of the patients. This study demonstrated that Nivolumab was safe and effective in pretreated gastric or GEJ cancer patients [64].

2 additional groups were added to the trial, who were treated with Nivolumab 1 mg/kg + Ipilimumab 3 mg/kg or Nivolumab 3 mg/kg + Ipilimumab 1 mg/kg every 3 weeks. Results showed that the TRAE of these 3 groups were almost identical [65].

The latest update on responses in the CHECKMATE-032 study is summarized below in Table **2**. Of the 160 patients who were enrolled in the three groups, 24% had PD-L1$^+$ ($\geq$1%) tumor cell. ORR was 19% in Nivolumab-alone group, 24% in N1+Ipilimumab 3 group, and 8% in N3+Ipilimumab 1 group, with overall higher ORR in PD-L1$^+$ ($\geq$1%) patients. In all patients, median overall survival was 6.2 months, 6.9 months and 4.8 months, respectively [66].

Table 2. Overall survival in all patients and patients with PD-L1 $\geq$1% in CHECKMATE-032.

All patients	Nivolumab alone n=59	N1+Ipilimumab 3 n=49	N3+Ipilimumab 1 N=52
Median overall survival (months)	6.2	6.9	4.8
OS Rate, %			
12 mo	39	35	24
18 mo	25	28	13
24 mo	22	22	-
Pts with PD-L1 $\geq$1%	n=16	n=10	n=13
Median overall survival (months)	6.2	NA	5.6

(Table 2) cont.....

All patients	Nivolumab alone n=59	N1+Ipilimumab 3 n=49	N3+Ipilimumab 1 N=52
OS Rate, %			
12 mo	34	50	23
18 mo	13	50	15

There is also a three arm trial comparing Nivolumab + Ipilimumab or Nivolumab + chemotherapy to chemotherapy alone in untreated advanced or metastatic gastric cancer patients, the results of which are awaited with interest, as this has the potential to move immunotherapy into first line treatment for this globally important cancer [66].

Another double blinded, randomized phase III clinical trial investigated Nivolumab as salvage treatment after second or later-line chemotherapy for advanced gastric cancer. 493 patients with gastric or GEJ cancer received3mg/kg Nivolumab (330) or placebo (163) every 2 weeks. The primary endpoint was overall survival (OS) and the data summarized in Table **3**, proved that Nivolumab is an effective salvage treatment for advanced gastric cancer [67].

Table 3. Results of Nivolumab (ONO-4538/BMS-936558) as salvage treatment after second or later-line chemotherapy for advanced gastric or GEJ cancer.

	Nivolumab	Placebo
Median OS	5.32 months	4.14 months
OS rate at 6 month	46.4%	34.7%
OS rate at 12 month	26.6%	10.9
ORR	11.2%	0%
Median PFS	1.61 months	1.45 months
Grade≥3 DRAEs	11.5%	5.5%

Colorectal Cancer

Deficient mismatch repair (dMMR or microsatellite instability, MSI) is one of the key genetic mechanisms driving the occurrence and progression of colorectal cancer (CRC). There are several genes controlling DNA mismatch repair function including MSH2, MLH1. One consequence of dMMR is that these tumour cells carry a very high neoantigen load due to the high frequency of mutations, increasing the likelihood of immune recognition. Perhaps unsurprisingly, microsatellite unstable colon tumours appear have a strong lymphocyte infiltration and have a significantly better prognosis than their microsatellite stable (MSS) counterparts, especially those diagnosed as stage II disease.

Takemoto *et al* showed that stroma-infiltrating lymphocytes (SIL) were found in approximately the same number in high grade microsatellite instability (MSI-H) patients (20%) and low grade microsatellite instability (MSI-L) or MSS tumors (12.8%). However, significant differences of intra-tumor cell-infiltrating lymphocytes (ITCIL) were shown between MSI-H CRC and MSI-L or MSS CRC patients (41.7% vs 4.3%, respectively (P< 0.001)). Furthermore, the prognosis of the tumors with higher ITCIL counts was better than the less infiltrated ones. In addition, increased PD-L1 expression has been found at the invasive edge of MSI-H tumors.

All the characteristics mentioned above, as well as the recent definition of highly immunogenic neo-antigens expressed in MSI-H tumor cells, suggest that MSI-H CRCs induce a protective host immune response that may reduce the incidence of metastasis formation and which might explain the better prognosis and potential responsiveness to immune checkpoint inhibition in this patient group [68].

In the initial 2 clinical trials conducted by Topalian *et al* patients with CRC were enrolled into both studies. In the total of 207 patients, there were 18 CRC patients and notably, only 1 patient with a metastatic-CRC, with PD-L1 positive tumor showed a complete response after 6 months' treatment of Nivolumab and had no signs of tumor recurrence after 3 years. This patient's tumor's genetic condition was found to be microsatellite unstable [69 - 71].

Since somatic mutations have the potential to encode "non-self" immunogenic antigens, it was hypothesized that tumors with mismatch-repair deficiency (dMMR) that can lead to thousands of somatic mutations, may be responsive to immune checkpoint inhibitors. Forty one patients with or without mismatch repair deficient advanced cancer were recruited into a phase II clinical trial of Pembrolizumab conducted by Le *et al*. Patients were treated with Pembrolizumab intravenously 10mg/kg every 2 weeks. They were separated into 3 cohorts of MMR-deficient CRC, MMR-proficient CRC and MMR-deficient non-colorectal cancer. The primary end point for the first two cohorts were the immune-related ORR and the immune-related PFS rate at 20 weeks. The primary endpoint for the third cohort was the immune-related PFS rate at 20 weeks.

The results showed that the 2 groups with MMR-deficient colorectal or non-colorectal cancer had the higher rate of immune-related OR (40% and 71%) and immune-related PFS at 20 weeks (78% and 67%) compared to the MMR-proficient colorectal cancer group (0% and 11%).

Interestingly, 1782 versus 73 somatic mutations per tumor in mismatch repair-deficient tumors and mismatch repair-proficient tumors was demonstrated (P=0.007), and higher somatic mutation loads were associated with prolonged PFS (P=0.02). Rash/pruritus, pancreatitis, and thyroiditis/hypothyroidism were found to be the most common TRAE, occurring in approximately 10% of patients. This phase II clinical trial proved that patients with mismatch-repair deficient tumors associated with a heavy load of somatic mutations, respond to anti-PD1 therapy [72]. The same investigators went on to demonstrate that a responding patient demonstrated rapid in vivo expansion of neoantigen-specific T cell clones that were reactive to mutant neopeptides found in the tumor, supporting the previously stated hypothesis that neoantigen load drives the response to immune checkpoint inhibition across a range of gastrointestinal and pancreatico-hepatobiliary tumours [73].

A recent update showed that of the 28 patients with mismatch repair deficiency, the progression free survival (PFS rate) was 61% and the overall survival (OS) rate was 66% at 24 months allowing them to conclude that patients with dMMR CRC receive durable clinical benefit from treatment with Pembrolizumab [74].

An open-label, multicenter, phase II study focusing on Nivolumab (3mg/kg Nivolumab every 2 weeks) in metastatic MSI-H colorectal cancer has been reported recently. The primary endpoint was investigator-assessed objective response rates and of the 74 patients who were able to be evaluated at present, 23 achieved an investigator assessed objective response suggesting that Nivolumab is an effective treatment for MSI-H metastatic CRC patients [75].

Also in the same study, a group of patients were treated with Nivolumab in combination with Ipilimumab therapy in MSH-H metastatic colorectal cancer. 3mg/kg Nivolumab plus 1mg/kg Ipilimumab was administered to the patients every 3 weeks for 4 cycles, followed by 3mg/kg Nivolumab alone every 2 weeks. Results were also encouraging that of the 27 patients enrolled, an ORR of 41% and DCR was 78% with acceptable safety profile was recorded [76]. Unlike MSI-H tumours, more research is needed to elucidate the effects of combination therapy on microsatellite stable colorectal cancer due to the extremely poor response in other trials of monotherapy [77].

Pancreatic Cancer

Pancreatic ductal adenocarcinoma is a poorly immunogenic malignant tumor with disappointing prognosis. Similar to colorectal cancer, the genetic condition of microsatellite instability of pancreatic cancer is found in around 2% of patients.

Le *et al*. conducted a phase II study to evaluate the activity of Pembrolizumab in mismatch repair deficient non-colorectal cancer. Of the first 17 patients with

MMR-deficient GI cancer enrolled, 4 patients had pancreatic cancer. 10 of the 17 non-colorectal GI cancer patients were evaluable. Half of whom responded (ORR was 50% and DCR was 70%) and who had an OS of 21 months. This study showed promising activity of Pembrolizumab in mismatch repair deficient GI cancers [78]. Another single-institution data showed increased response in mismatch repair deficient gastrointestinal cancers including 2 pancreatic cancer patients [79]. Finally, in 2017, FDA approved the use of Pembrolizumab or Nivolumab in advanced cancer patients, with tumours of any origin with mismatch repair deficiency, thus identifying MMRD as the first pan tumour companion diagnostic.

A phase II clinical trial was designed to assess the efficacy of Ipilimumab (an anti-CTLA-4 antibody) in locally advanced or metastatic pancreatic adenocarcinoma. 27 patients were given 3.0mg/kg Ipilimumab every 3 weeks, however, no significant responses were observed [80].

2 years later after the publication of anti-CTLA-4 monotherapy in pancreatic cancer, a phase 1 study including multiple cancer types showed that no response were witnessed in the phase I study of the anti-PD-L1 antibody Nivolumab monotherapy in 14 pancreatic cancer patients [81].

Since preclinical research indicated the prospect of synergy between anti-CTLA-4 antibodies and granulocyte macrophage colony stimulating factor (GM-CSF) cell-based vaccines in some cancer types [82 - 84] more efforts have been made to evaluate the activity of combination therapy involving immune checkpoint antibodies.

Le *et al* enrolled 30 pretreated pancreatic ductal adenocarcinoma patients into 2 groups (1:1) who were randomized to 10mg/kg Ipilimumab alone or 10mg/kg Ipilimumab plus GVAX. Results showed that 3 patients in the combination group had prolonged disease stabilization and 7 patients had a decline in the plasma level of the tumour biomarker CA19-9. Median OS was 3.6 versus 5.7 months and 1-year OS was 7% versus 27%, both favoring the combination group [85].

Combining the anti CTLA4 antibody Tremelimumab with gemcitabine in metastatic pancreatic cancer patients has been investigated in a phase Ib trial. 34 patients who been administered chemotherapy were treated with an escalating dose of Tremelimumab (with maximum of 15 mg/kg). Median survival was 7.4 months and only 1 patient experienced severe TRAE (diarrhea with dehydration), however, only 2 patients had partial responses [86].

More recently, a similar phase Ib study assessed the safety and activity of Ipilimumab plus gemcitabine and showed that 12 out of 16 patients discontinued the study because of progression of disease after 1 cycle of treatment while 2 developed PR whilst no severe DRAEs were observed [87].

These studies above demonstrated the relative safety of combining immune checkpoint inhibitors with gemcitabine in chemotherapy-naïve pancreatic cancer patients, but whether it can significantly improve the responsiveness needs further exploration in randomized settings.

Durvalumab, another anti-PD-L1 monoclonal antibody, is now undergoing trials combined with the promising Bruton's tyrosine kinase inhibitor Ibrutinib [88].

HISTONE DEACETYLASE INHIBITORS

Post-translational modification of a gene describes the epigenetic process whereby gene expression is regulated beyond the transcription and translation phase. One of the ways in which this can occur is via alterations to histone tails by phosphorylation, acetylation or methylation, which in turn results in either increased expression or silencing of genes. Briefly, histones form part of a nucleosome complex, around which DNA is tightly bound to create chromatin strands, which themselves are twisted and condensed to form chromosomes. This nucleosome complex consists of two copies of the histones H2B, H2A, H4 and H3 resulting in an octamer [89]. The histones possess positively charged tails which tightly interact with the bound DNA decreasing accessibility of promotor regions to transcription factors. Acetylation of the tails neutralizes their positive charge, weakening the bond and exposing gene promotor regions on the DNA strands which then become accessible to transcription machinery. Histone deacetylases (HDAC) remove acetyl groups from histone tails, restoring their positive charge and their subsequent strong interaction with the DNA double helix, which ultimately maintains the structure of the chromatin strands but simultaneously prevents the recruitment of transcription factors to promoter regions [90] and this ultimately leads to gene silencing.

HDACs are a heterogenous group of enzymes, capable of both histone and non-histone protein acetylation and are categorized according to their homology to yeast, their location within a cell and their enzymatic activity. Class I HDACs have a similar homology to RDP3 yeast, are ubiquitously expressed and mostly concentrated within the nucleus. Both Class IIa and Class IIb are homologous to Hda1 yeast protein and their expression is tissue specific however, Class IIa is more mobile, shuttling between the nucleus and the cytoplasm, whereas Class IIb is primarily located in the cytoplasm and has a low enzymatic activity in comparison. Class IV is tissue specific however it's homology to yeast is unknown. These represent the classical HDACs which are sensitive to HDAC inhibitors (HDACi). A third class, Class III, also known as sirtuins has a homology to yeast Sir2, however it is not sensitive to HDACi's [91].

Abnormal expression of HDAC has been implicated in tumorigenesis and indeed studies have shown that HDACs are over expressed in prostate cancer [92] and gastrointestinal cancers [93, 94] and their presence has been associated with poorer prognosis.

HDAC inhibitors have been shown to have anti-tumour effects, however their exact mode of action is yet to be fully elucidated but multiple mechanisms have been proposed which have been summarized in 2 recent review articles [95, 96]. These include the arrest of the tumour cell cycle in the G1/G2 phase; initiation of apoptosis in the tumour cell via activation of the intrinsic and extrinsic pathways; inhibition of angiogenesis, metastasis and tumour invasion; and increasing tumour immunogenicity. To further explain, G1 cell cycle arrest has been attributed to increased transcription of the p21 gene as a result of released HDAC mediated suppression by HDAC inhibitors. P21 gene encodes p21 protein, a cyclin dependent kinase inhibitor (CDKI) which regulates cell cycle progression and prevents uncontrolled proliferation. However, the mechanism for G2 cell cycle arrest is at present poorly understood. HDAC inhibitor mediated activation of apoptosis is *via* the upregulation of death receptors (such as TNF receptors) in the extrinsic pathway, and the increased transcription of pro-apoptotic proteins (such as Bim and Bmf which belong to the BCL2 family of pro-apoptotic proteins) in the intrinsic pathway. Pro-angiogenic factors such as vascular endothelial growth factor (VEGF) and hypoxia inducible factor 1α (HIF1α) are decreased in the presence of HDAC inhibitors. This, along with the reduced expression of the chemokine receptor CXCR4, which directs endothelial cells to sites of angiogenesis, is a mechanism by which HDAC inhibitors prevent neovascularization and in doing so indirectly impair metastases. Finally, HDAC inhibitors have been shown to enhance tumour immunogenicity by upregulating MHC complexes and co-stimulatory molecules making them more visible to the host immune system.

Clinical Trials of Histone Deacetylase Inhibitors in Gastrointestinal and Pancreatic Cancers

Gastrointestinal Cancer

Like HDAC, HDACi represent a varied group of inhibitors classified according to their chemical structure, of which there are 4: hydroxamic acids, cyclic peptides, benzamides and short-chain fatty acids [97, 98]. At present there are only 3 HDAC inhibitors approved by the FDA for clinical use and these include vorinostat, romidepsin and belinostat, all of which are used for treating haematological malignancies [98]. Although there are approximately 300 clinical trials either completed or ongoing [99, 100], investigating the use of different classes of HDACI's, either as monotherapy or in combination with traditional anti-tumour agents in both solid and haematological malignancies, most of these trials are in the phase I or phase II stages and results have been mixed. Unlike in haematological malignancies, oral vorinostat has not shown much success in the treatment of gastrointestinal cancers. A phase I trial showed promising results when used as monotherapy at a dose of 300mg in patients with gastric, colon, and rectal cancer. No dose limiting toxicity was reported at this dose and stable disease for a period of more than 8 weeks was observed in some of the study subjects [101].

However, oral vorinostat has not been as successful when used in combination with traditional chemotherapy agents. A phase II trial looking into the effect of two doses of oral vorinostat (high dose *vs* low dose) given in combination with 5-fluorouracil (5-FU) to patients with refractory colorectal cancer did not demonstrate any significant clinical benefit and therefore did not warrant further investigation [102]. Inability to establish maximum tolerated dose due to dose limiting toxicities at all dose levels resulted in the early termination of another phase I/II trial which was conducted to investigate dual therapy of 5-fluorouracil and oral vorinostat in patients with metastatic colorectal cancer [103].

Similarly disappointing findings have also been reported in the context of gastric cancers. An initial phase I trial on capecitabine and cisplatin combined with oral vorinostat was performed in order to identify recommended phase II trial dose. Initial results of this phase 1 combination study were promising and the authors identified a recommended phase II dose of 400mg of oral vorinostat which yielded an objective response rate, median progression free survival and overall survival of 56%, 7.1 months and 18 months respectively. This dose was also associated with the highest levels of H3 acetylation which is an indicator of HDACi activity [104]. Unfortunately, a follow up phase II trial did not show any clinical efficacy as the study failed to meet its primary end point of a 6-month progression free survival rate of 60% and in addition higher levels of toxicities were reported compared to traditional cytotoxic regimes [105].

Despite these disappointing results, there still remains an appetite for further exploration of the use of HDACis to treat gastrointestinal cancers not least because preclinical studies continue to report positive anti-tumour activity in human cell lines albeit with different types of drugs. Granted that so far, no trial has progressed to the phase III stage, this is largely in part due to an imbalance between clear clinical benefit and acceptable toxicity. This however, should not negate the pursuit of translating preclinical success into clinical benefit. Indeed, different classes of HDACis provide different toxicity profiles and potentially differing modes of action. A recent preclinical study showed that the HDACi depsipeptide, when used in combination with 5-FU in human colon cancer cell lines resulted in upregulation of MHC II and p21 and activated caspase 3/7 which was associated with a reduction in tumour colony formation [106]. Furthermore, CHR-3996, which is a selective HDAC inhibitor has shown evidence of growth inhibition of various human cell lines in the preclinical setting [107]. Encouraged by the laboratory findings of the latter, a phase I study, looking into the pharmacokinetics and pharmacodynamics of CHR-3995 as monotherapy in refractory solid tumours (the majority of which were from the upper gastrointestinal, lower gastrointestinal, pancreatic and hepatobiliary system) was recently conducted. The authors were able to identify a recommended phase II dose of 40mg once day, a dose at which both acceptable level of toxicity and the highest level of H3 acetylation were achieved. They also reported clinical efficacy, with 9 out of 39 patients showing stable disease and 1 patient with confirmed partial response [108]. These results, although modest, provide enticing evidence for the clinical benefits of using different classes of HDACi in the treatment of gastrointestinal cancers. It remains to be seen if similar encouraging results will be reported in phase II trials. Indeed, the results of a recently completed phase I/II trial conducted to investigate the safety profile and efficacy of resminostat combined with irinotecan, fluorouracil, and folinic acid chemotherapy regime (FOLFIRI) in patients with K-ras mutated advanced colorectal cancer, are yet to be published [109].

Table 4. Discussed clinical trials and their outcomes.

Ref	Time/Name	Phase	Number of Patient	Patient Condition	Method	Outcome
[55]	2010	II	18	Metastatic gastric and esophageal adenocarcinoma	Tremelimumab	1 patient achieved durable clinical benefit
[56, 57]	2015	Ib	20	Advanced gastric cancer	Avelumab	ORR 15%, DCR 65% PD-L1 expression in 19 patients, with Higher ORR and PFS rate.

(Table 4) cont.....

Ref	Time/Name	Phase	Number of Patient	Patient Condition	Method	Outcome
[58]	2016	II	151(62 as 2L VS 89 as 1L)	Gastric or GEJ cancer	Avelumab	15 with Grade ≥3 TRAE, 1 with TRD, DCR: 29% VS 57.3% Median PFS: 6 weeks VS 12 weeks PD-L1 expression assessed in 74 patients
[59]	2016 KEYNOTE-012	Ib	39	Recurrent or metastatic Gastric or GEJ adenocarcinoma	Pembrolizumab	36 met evaluable disease, OR: 22% 5 with Grade 3 or 4 TRAE
[60 61]	2016 KEYNOTE-059	II	259	advanced gastric or gastroesophageal junction (G/GEJ) cancer	Pem monotherapy VS Pem+5-FU+Cisplatin	17 with Grade ≥3 TRAE PD-L1 expression in 142 patients Overall ORR 11.2% CR 1.9%, PR 9.3%, SD 17%, PD 55.6%
[62] 2016	2017	Ib/II		Relapsed/ refractory advanced HER2+ GEJ or Gastric cancer	Pem and Margetuximab (Fc-optimized monoclonal Ab targets HER2)	Ongoing
[63]	2017	Ia/b	40	Pretreated Gastric /GEJ Adenocarcinoma	Ramucirumab (Anti-VEGF inhibitor) + Pem	The safety was well tolerated 7.5% have responded to the therapy.
[64]	2016 CHECKMATE-032	I/II	59	Advanced Gastric/ GEJ Cancer	Nivolumab	ORR 12%, Median OS: 7 mo, 12moth OSR 38% 7 with Grade 3 TRAE
[67]	2017	III	493	Gastric or GEJ Cancer	Nivo VS Placebo	Salvage Treatment

(Table 4) cont.....

Ref	Time/Name	Phase	Number of Patient	Patient Condition	Method	Outcome
[69-71]	2012		18	Colorectal Cancer	Nivolumab	1 met CR and no recurrence after 3 years This patient was microsatellite unstable.
[72, 74]	2015	II	41	Advanced Cancer	Pembrolizumab	Patients with MMR-deficiency CRC had higher rate of immune-related OR and immune-related PFS than MMR-proficient CRC. PFS rate 61%, OS rate 66% in MMR deficient cancer patients.
[75]	2017	II	74	Microsatellite Instability- High CRC	Nivolumab	23 achieved objective response
[76]	2017	II	27	Microsatellite Instability- High CRC	Nivolumab+ Ipilimumab (Anti-CTLA-4 Ab)	ORR 41% DCR 78% with acceptable safety profile
[77]	2016	II	17	MMR-deficient GI cancer including 4 Pancreatic Cancer	Pembrolizumab	ORR 50%; DCR 70%; 21-month OS; Showing promising activity of Pem in MMR-deficient GI cancer
[80]	2010	II	27	Locally advanced or metastatic pancreatic adenocarcinoma	Ipilimumab	No significant responses
[81]	2015	I	14	Pancreatic Cancer	Nivolumab	No response

(Table 4) cont.....

Ref	Time/Name	Phase	Number of Patient	Patient Condition	Method	Outcome
[85]	2013		30	Pretreated Pancreatic Ductal Adenocarcinoma	Ipilimumab VS Ipi+GVAX	In Ipi+GVAX group, 3 patients had prolonged SD and 7 had a decline of CA19-9
[86]	2014	Ib	34	Metastatic Pancreatic Cancer	Tremlimumab+ Gemcitabine	Median survival 7.4 months 1 patient met severe TRAE 2 patients had PR
[87]	2016	Ib	16	NA	Ipilimumab+ Gemcitabine	12 discontinued because of PD 2 developed PR No severe TRAE
[101]	2011	I	20	Gastric, Colon and Rectal Cancer	Vorinostat	No dose limiting toxicity was reported at dose of 300 mg and patients achieve 8 weeks SD period.
[102]	2006	II	58	Refractory colorectal cancer	Vorinostat+5-Fu (high dose vs low dose)	No significant clinical benefit
[104]	2014	I	30	Gastric Cancer	Capecitabine + Cisplatin +Vorinostat	Recommendation of 400mg oral vorinostat rate ORR 56%, Median PFS 7.1 mo, OS 18mo
[105]	2016	II	38	Gastric Cancer	Capecitabine + Cisplatin + Vorinostat	No clinical efficacy Higher levels of toxicities were noticed compared to traditional cytotoxic regimens

(Table 4) cont.....

Ref	Time/Name	Phase	Number of Patient	Patient Condition	Method	Outcome
[108]	2012	I	39	Refractory solid tumor, most of them are from the upper & lower intestinal, pancreatic and hepatobiliary system	CHR-3995	Identification of a recommended phase II doses of 40 mg with acceptable toxicity 9 achieved SD 1 met PR
[111]	2014	I/II	12	Advanced pancreatobiliary tract malignancy	Valproic acid + S-1	Feasible and well tolerated 1 met PR, 10 met SD, 1 met PD
[112]	2006	II	174	Liver, Lymph nodes, Lung, Peritoneum, Bowel, other	Gemcitabine+CI-994 VS Gemcitabine + Placebo	No significant clinical benefits were observed but with worse toxicity effects.

ORR: Objective Response Rate; DCR: Disease Control Rate; Mo: Month; 2L: Second-Line; 1L: First-Line; GEJ: Gastro-Esophageal Junction; TRAE: Treatment-Related Adverse Events; TRD: Treatment-Related Death; PFS: Progression Free Survival; OR: Overall Response; CR: Complete Response; PR: Partial Response; SD: Stable Disease; PD: Progressive Disease; VEGF: Vascular Endothelial Growth Factor; MMR: mismatch repair; CRC: Colorectal Cancer; DCR: Disease Control Rate;

Pancreatic Cancer

Pancreatic cancer remains a terminal diagnosis, the treatment for which is very limited, therefore, clinical trials investigating novel therapies to improve prognosis are welcomed. HDACis have become more visible in the pancreatic cancer landscape, in part due to the success of anti-tumour activity in the preclinical setting. An example would be the observed enhancement of the anti-tumour effects of 5-FU by valproic acid, a short chain fatty acid HDAC inhibitor, *in vitro*. The authors of the study in which this observation was identified showed that proliferation of pancreatic cell lines was significantly impaired when exposed to both 5-FU and valproic acid than when either was used alone (p<0.01) [110]. Encouraged by these findings, a phase I/II trial to investigate the safety and efficacy of combination therapy with valproic acid and S-1 (an oral flouropyrimidine derivative) in patients with advanced pancreatobiliary tract malignancies subsequently followed. The study reported that this combination was both feasible and well tolerated and of the 12 subjects enrolled, 1 patient had a partial response, 10 had stable disease and 1 showed progressive disease [111].

Combination of HDACi with gemcitabine has not been as positive. A randomized phase II double blind, placebo controlled multicentre study failed to show any improvement in overall survival in patients who received gemcitabine + CI-994 (an oral histone deacetylase inhibitor) *versus* those who received gemcitabine + placebo. Furthermore, haematological toxicity was more pronounced in the former and the authors concluded that this combination was inferior to gemcitabine alone [112]. However, a recent preclinical study has renewed hope for the use of HDACi with gemcitabine. CG200745, a pan-HDACi inhibitor similar to vorinostat, inhibited pancreatic cancer cell growth and synergism was observed when combined with gemcitabine/erlotinib. Furthermore, the authors observed that pancreatic cell lines resistant to gemcitabine were re-sensitized once exposed to CG200745 and eventually succumbed to the anti-proliferative effects of the cytotoxic agents [113].

Were this to translate into the clinical setting, it would provide a much needed and very exciting alternative to treat a currently insurmountable biological hurdle: the development of resistance to gemcitabine. Certainly, therefore, efforts should be made to further explore the feasibility of this combination in patients with pancreatic cancer, in the hope that a positive result will provide more treatment options for what is an already devastating diagnosis.

HR23B: A Marker of CXD101 Sensitivity

A genome-wide loss-of-function screen has identified HR23B, a protein that shuttles ubiquitinated cargo proteins to the proteasome for degradation as a sensitivity determinant for HDAC inhibitor-induced cell apoptosis [114]. Proteasome activity is deregulated by HDACi through a HR23B dependent mechanism, and HDACi sensitize Cutaneous T Cell Lymphoma (CTCL) cells to the effects of proteasome inhibitors. Khan and colleagues [115] evaluated the role of HR23B in predicting response to HDACi within CTCL cells by evaluating HR23B expression in CTCL cell lines. HR23B expression influenced sensitivity to HDACi-induced apoptosis and specific manipulation of HR23B levels within CTCL cells in vitro altered HDACi sensitivity. In another study, samples of CTCL were collected following a Phase II trial of the HDAC inhibitor romidepsin in CTCL [116]. High HR23B expression by immunohistochemistry (IHC) in this series of relapsed CTCL *in-situ* correlated with favourable response to HDAC inhibition. HR23B expression in 65 CTCL biopsies had a positive predictive value (PPV) of 71.7% [115]. These results suggest proteasome deregulation contributes to the anti-cancer activity of HDACi and raises the hypothesis that HR23B could provide a biomarker for identifying tumours predicted to respond favourably to HDACi.

Further evidence suggesting HR23B has potential as a relevant biomarker is from published pre-clinical studies in hepatocellular carcinoma (HCC) and medulloblastoma. A novel cell line HD-MB03 was isolated from a patient with metastatic Group 3 medulloblastoma. Analysis of protein expression from HD-MB03 revealed intermediate to strong expression of HDAC's 2, 5, 8, and 9 and the predictive marker HR23B. Treatment with the HDAC inhibitors helminthosporium carbonum (HC)-toxin, vorinostat, and panobinostat resulted in high levels of cell death and augmented radiation-sensitivity [117]. In patients with unresectable HCC and chronic liver disease, belinostat was assessed in a phase I/II study [118]. A secondary exploratory analysis revealed tumours from patients with stable disease (SD) (CR plus PR plus SD) post-treatment, contained high and low HR23B IHC scores in 58% and 14%, respectively (p=0.036), confirming the positive selective advantage conferred by HR23B patient stratification.

Can We Reverse Immunoediting by Treating with Histone Deacetylase (HDAC) Inhibitors?

Researchers have hypothesized that strategies which increase expression of T-cell chemokines and T-cell infiltration of tumors would be capable of enhancing response to PD-1 blockade. There is evidence to suggest that histone deacetylase inhibitors (HDACi) [119, 120] are capable of inducing expression of these chemokines in tumor and increasing immune recognition. It has been reported that increased histone acetylation induced by HDAC inhibitors results in the increase expression of MHC molecules and other molecules involved in antigen processing and presentation [121 - 124]. Also it can increase expression of tumor antigens recognized by cytotoxic T lymphocytes (CTLs) and ligands for NK activating receptors [125, 126]. The HDACi romidepsin, induced a strong antitumor response against KRAS mutant NSCLC tumors in mice which correlated with T cell infiltration of the tumor, and CD8 T-cell infiltration in human lung tumors has been shown to increase after HDACi vorinostat treatment [127, 128]. The combination of the HDAC I depsipeptide and very low concentrations of the cytotoxic antimetabolite 5-fluorouracil (5-FU) induces apoptosis synergistically and up regulates MHC class II in human colon cancer HCT-116 cells [106]. Based on these and many other preclinical study results, several clinical trials have been initiated to evaluate whether the combination of HDAC inhibitors and anti-PD1 therapies can improve tumor responses by enhancing the CD8 T cell infiltration. (NCT02638090, NCT02437136, NCT02697630, NCT01928576, NCT024353620 and NCT02708680) These trials may give an indication if HDAC inhibitors can improve response to anti-PD1 agents in the coming future.

CONCLUSION

As mentioned in this article, it is the loss of function of cellular immune system, that complies possible reason causing no response to checkpoint blockade therapy in colorectal cancer. During the cell deterioration process of CRC, the deficiency of immuno stimulatory signal presentation and the activation of immunological checkpoints may suppress the immunosurveillance [129]. One clinical approach to targeting the mechanisms that underlie immune cancer may be to combine immune checkpoint and HDAC inhibitors to restore immunoreactivity and enhance tumor cell kill.

CONSENT FOR PUBLICATION

Not applicable.

CONFLICT OF INTEREST

DJ Kerr reports that he is a Director of Celleron Therapeutics

ACKNOWLEDGEMENTS

Declared none.

REFERENCES

[1]   Hanahan D, Weinberg RA. Hallmarks of cancer: the next generation. Cell 2011; 144(5): 646-74.
        [http://dx.doi.org/10.1016/j.cell.2011.02.013] [PMID: 21376230]

[2]   Bonilla FA, Oettgen HC. Adaptive immunity. J Allergy Clin Immunol 2010; 125(2) (Suppl. 2): S33-40.
        [http://dx.doi.org/10.1016/j.jaci.2009.09.017] [PMID: 20061006]

[3]   Chen L, Flies DB. Molecular mechanisms of T cell co-stimulation and co-inhibition. Nat Rev Immunol 2013; 13(4): 227-42.
        [http://dx.doi.org/10.1038/nri3405] [PMID: 23470321]

[4]   Miyara M, Sakaguchi S. Natural regulatory T cells: mechanisms of suppression. Trends Mol Med 2007; 13(3): 108-16.
        [http://dx.doi.org/10.1016/j.molmed.2007.01.003] [PMID: 17257897]

[5]   Sharpe AH, Freeman GJ. The B7-CD28 superfamily. Nat Rev Immunol 2002; 2(2): 116-26.
        [http://dx.doi.org/10.1038/nri727] [PMID: 11910893]

[6]   Dunn GP, Bruce AT, Ikeda H, Old LJ, Schreiber RD. Cancer immunoediting: from immunosurveillance to tumor escape. Nat Immunol 2002; 3(11): 991-8.
        [http://dx.doi.org/10.1038/ni1102-991] [PMID: 12407406]

[7]   Schreiber RD, Old LJ, Smyth MJ. Cancer immunoediting: integrating immunity's roles in cancer suppression and promotion. Science 2011; 331(6024): 1565-70.
        [http://dx.doi.org/10.1126/science.1203486] [PMID: 21436444]

[8]   Koebel CM, Vermi W, Swann JB, et al. Adaptive immunity maintains occult cancer in an equilibrium state. Nature 2007; 450(7171): 903-7.
        [http://dx.doi.org/10.1038/nature06309] [PMID: 18026089]

[9]   Teng MW, Vesely MD, Duret H, *et al.* Opposing roles for IL-23 and IL-12 in maintaining occult cancer in an equilibrium state. Cancer Res 2012; 72(16): 3987-96.
[http://dx.doi.org/10.1158/0008-5472.CAN-12-1337] [PMID: 22869585]

[10]   Beyersdorf N, Kerkau T, Hünig T. CD28 co-stimulation in T-cell homeostasis: a recent perspective. ImmunoTargets Ther 2015; 4: 111-22.
[PMID: 27471717]

[11]   Nagaraj S, Gabrilovich DI. Tumor escape mechanism governed by myeloid-derived suppressor cells. Cancer Res 2008; 68(8): 2561-3.
[http://dx.doi.org/10.1158/0008-5472.CAN-07-6229] [PMID: 18413722]

[12]   Wolfram RM, Budinsky AC, Brodowicz T, *et al.* Defective antigen presentation resulting from impaired expression of costimulatory molecules in breast cancer. Int J Cancer 2000; 88(2): 239-44.
[http://dx.doi.org/10.1002/1097-0215(20001015)88:2<239::AID-IJC15>3.0.CO;2-Z]   [PMID: 11004675]

[13]   Thomas GR, Chen Z, Leukinova E, Van Waes C, Wen J. Cytokines IL-1 alpha, IL-6, and GM-CSF constitutively secreted by oral squamous carcinoma induce down-regulation of CD80 costimulatory molecule expression: restoration by interferon gamma. Cancer Immunol Immunother 2004; 53(1): 33-40.
[http://dx.doi.org/10.1007/s00262-003-0433-4] [PMID: 14551747]

[14]   Ugurel S, Uhlig D, Pföhler C, Tilgen W, Schadendorf D, Reinhold U. Down-regulation of HLA class II and costimulatory CD86/B7-2 on circulating monocytes from melanoma patients. Cancer Immunol Immunother 2004; 53(6): 551-9.
[http://dx.doi.org/10.1007/s00262-003-0489-1] [PMID: 14727087]

[15]   Lindauer M, Rudy W, Gückel B, Doeberitz MV, Meuer SC, Moebius U. Gene transfer of costimulatory molecules into a human colorectal cancer cell line: requirement of CD54, CD80 and class II MHC expression for enhanced immunogenicity. Immunology 1998; 93(3): 390-7.
[http://dx.doi.org/10.1046/j.1365-2567.1998.00450.x] [PMID: 9640250]

[16]   Jin HT, Anderson AC, Tan WG, *et al.* Cooperation of Tim-3 and PD-1 in CD8 T-cell exhaustion during chronic viral infection. Proc Natl Acad Sci USA 2010; 107(33): 14733-8.
[http://dx.doi.org/10.1073/pnas.1009731107] [PMID: 20679213]

[17]   Sakuishi K, Apetoh L, Sullivan JM, Blazar BR, Kuchroo VK, Anderson AC. Targeting Tim-3 and PD-1 pathways to reverse T cell exhaustion and restore anti-tumor immunity. J Exp Med 2010; 207(10): 2187-94.
[http://dx.doi.org/10.1084/jem.20100643] [PMID: 20819927]

[18]   Nishimura H, Nose M, Hiai H, Minato N, Honjo T. Development of lupus-like autoimmune diseases by disruption of the PD-1 gene encoding an ITIM motif-carrying immunoreceptor. Immunity 1999; 11(2): 141-51.
[http://dx.doi.org/10.1016/S1074-7613(00)80089-8] [PMID: 10485649]

[19]   Sica GL, Choi IH, Zhu G, *et al.* B7-H4, a molecule of the B7 family, negatively regulates T cell immunity. Immunity 2003; 18(6): 849-61.
[http://dx.doi.org/10.1016/S1074-7613(03)00152-3] [PMID: 12818165]

[20]   Chen LJ, Sun J, Wu HY, *et al.* B7-H4 expression associates with cancer progression and predicts patient's survival in human esophageal squamous cell carcinoma. Cancer Immunol Immunother 2011; 60(7): 1047-55.
[http://dx.doi.org/10.1007/s00262-011-1017-3] [PMID: 21519829]

[21]   Arigami T, Uenosono Y, Ishigami S, Hagihara T, Haraguchi N, Natsugoe S. Clinical significance of the B7-H4 coregulatory molecule as a novel prognostic marker in gastric cancer. World J Surg 2011; 35(9): 2051-7.
[http://dx.doi.org/10.1007/s00268-011-1186-4] [PMID: 21748517]

[22]   Jiang J, Zhu Y, Wu C, *et al.* Tumor expression of B7-H4 predicts poor survival of patients suffering from gastric cancer. Cancer Immunol Immunother 2010; 59(11): 1707-14.
[http://dx.doi.org/10.1007/s00262-010-0900-7] [PMID: 20725832]

[23]   Awadallah NS, Shroyer KR, Langer DA, *et al.* Detection of B7-H4 and p53 in pancreatic cancer: potential role as a cytological diagnostic adjunct. Pancreas 2008; 36(2): 200-6.
[http://dx.doi.org/10.1097/MPA.0b013e318150e4e0] [PMID: 18376314]

[24]   Xu H, Chen X, Tao M, *et al.* B7-H3 and B7-H4 are independent predictors of a poor prognosis in patients with pancreatic cancer. Oncol Lett 2016; 11(3): 1841-6.
[http://dx.doi.org/10.3892/ol.2016.4128] [PMID: 26998087]

[25]   Shiraki K, Tsuji N, Shioda T, Isselbacher KJ, Takahashi H. Expression of Fas ligand in liver metastases of human colonic adenocarcinomas. Proc Natl Acad Sci USA 1997; 94(12): 6420-5.
[http://dx.doi.org/10.1073/pnas.94.12.6420] [PMID: 9177233]

[26]   Ljunggren HG, Kärre K. In search of the 'missing self': MHC molecules and NK cell recognition. Immunol Today 1990; 11(7): 237-44.
[http://dx.doi.org/10.1016/0167-5699(90)90097-S] [PMID: 2201309]

[27]   Smyrk TC, Watson P, Kaul K, Lynch HT. Tumor-infiltrating lymphocytes are a marker for microsatellite instability in colorectal carcinoma. Cancer 2001; 91(12): 2417-22.
[http://dx.doi.org/10.1002/1097-0142(20010615)91:12<2417::AID-CNCR1276>3.0.CO;2-U] [PMID: 11413533]

[28]   De Smedt L, Lemahieu J, Palmans S, *et al.* Microsatellite instable vs stable colon carcinomas: analysis of tumour heterogeneity, inflammation and angiogenesis. Br J Cancer 2015; 113(3): 500-9.
[http://dx.doi.org/10.1038/bjc.2015.213] [PMID: 26068398]

[29]   Cabrera CM, Jiménez P, Cabrera T, Esparza C, Ruiz-Cabello F, Garrido F. Total loss of MHC class I in colorectal tumors can be explained by two molecular pathways: beta2-microglobulin inactivation in MSI-positive tumors and LMP7/TAP2 downregulation in MSI-negative tumors. Tissue Antigens 2003; 61(3): 211-9.
[http://dx.doi.org/10.1034/j.1399-0039.2003.00020.x] [PMID: 12694570]

[30]   Watson NF, Ramage JM, Madjd Z, *et al.* Immunosurveillance is active in colorectal cancer as downregulation but not complete loss of MHC class I expression correlates with a poor prognosis. Int J Cancer 2006; 118(1): 6-10.
[http://dx.doi.org/10.1002/ijc.21303] [PMID: 16003753]

[31]   Colegio OR, Chu NQ, Szabo AL, *et al.* Functional polarization of tumour-associated macrophages by tumour-derived lactic acid. Nature 2014; 513(7519): 559-63.
[http://dx.doi.org/10.1038/nature13490] [PMID: 25043024]

[32]   Busuttil RA, George J, Tothill RW, *et al.* A signature predicting poor prognosis in gastric and ovarian cancer represents a coordinated macrophage and stromal response. Clin Cancer Res 2014; 20(10): 2761-72.
[http://dx.doi.org/10.1158/1078-0432.CCR-13-3049] [PMID: 24658156]

[33]   Curiel TJ, Coukos G, Zou L, *et al.* Specific recruitment of regulatory T cells in ovarian carcinoma fosters immune privilege and predicts reduced survival. Nat Med 2004; 10(9): 942-9.
[http://dx.doi.org/10.1038/nm1093] [PMID: 15322536]

[34]   Wolf D, Wolf AM, Rumpold H, *et al.* The expression of the regulatory T cell-specific forkhead box transcription factor FoxP3 is associated with poor prognosis in ovarian cancer. Clin Cancer Res 2005; 11(23): 8326-31.
[http://dx.doi.org/10.1158/1078-0432.CCR-05-1244] [PMID: 16322292]

[35]   Svensson H, Olofsson V, Lundin S, *et al.* Accumulation of CCR4⁺CTLA-4 FOXP3⁺CD25(hi) regulatory T cells in colon adenocarcinomas correlate to reduced activation of conventional T cells. PLoS One 2012; 7(2): e30695.

[http://dx.doi.org/10.1371/journal.pone.0030695] [PMID: 22319577]

[36] Brudvik KW, Henjum K, Aandahl EM, Bjørnbeth BA, Taskén K. Regulatory T-cell-mediated inhibition of antitumor immune responses is associated with clinical outcome in patients with liver metastasis from colorectal cancer. Cancer Immunol Immunother 2012; 61(7): 1045-53.
[http://dx.doi.org/10.1007/s00262-011-1174-4] [PMID: 22159472]

[37] Salama P, Phillips M, Grieu F, *et al.* Tumor-infiltrating FOXP3+ T regulatory cells show strong prognostic significance in colorectal cancer. J Clin Oncol 2009; 27(2): 186-92.
[http://dx.doi.org/10.1200/JCO.2008.18.7229] [PMID: 19064967]

[38] Savage PA, Malchow S, Leventhal DS. Basic principles of tumor-associated regulatory T cell biology. Trends Immunol 2013; 34(1): 33-40.
[http://dx.doi.org/10.1016/j.it.2012.08.005] [PMID: 22999714]

[39] Liu J, Zhang N, Li Q, *et al.* Tumor-associated macrophages recruit CCR6+ regulatory T cells and promote the development of colorectal cancer via enhancing CCL20 production in mice. PLoS One 2011; 6(4): e19495.
[http://dx.doi.org/10.1371/journal.pone.0019495] [PMID: 21559338]

[40] Geng Y, Wang H, Lu C, *et al.* Expression of costimulatory molecules B7-H1, B7-H4 and Foxp3+ Tregs in gastric cancer and its clinical significance. Int J Clin Oncol 2015; 20(2): 273-81.
[http://dx.doi.org/10.1007/s10147-014-0701-7] [PMID: 24804867]

[41] Hiraoka N, Onozato K, Kosuge T, Hirohashi S. Prevalence of FOXP3+ regulatory T cells increases during the progression of pancreatic ductal adenocarcinoma and its premalignant lesions. Clin Cancer Res 2006; 12(18): 5423-34.
[http://dx.doi.org/10.1158/1078-0432.CCR-06-0369] [PMID: 17000676]

[42] Chen L. Co-inhibitory molecules of the B7-CD28 family in the control of T-cell immunity. Nat Rev Immunol 2004; 4(5): 336-47.
[http://dx.doi.org/10.1038/nri1349] [PMID: 15122199]

[43] Linsley PS, Brady W, Urnes M, Grosmaire LS, Damle NK, Ledbetter JA. CTLA-4 is a second receptor for the B cell activation antigen B7. J Exp Med 1991; 174(3): 561-9.
[http://dx.doi.org/10.1084/jem.174.3.561] [PMID: 1714933]

[44] Chen DS, Mellman I. Oncology meets immunology: the cancer-immunity cycle. Immunity 2013; 39(1): 1-10.
[http://dx.doi.org/10.1016/j.immuni.2013.07.012] [PMID: 23890059]

[45] Wing K, Onishi Y, Prieto-Martin P, *et al.* CTLA-4 control over Foxp3+ regulatory T cell function. Science 2008; 322(5899): 271-5.
[http://dx.doi.org/10.1126/science.1160062] [PMID: 18845758]

[46] Lenschow DJ, Walunas TL, Bluestone JA. CD28/B7 system of T cell costimulation. Annu Rev Immunol 1996; 14: 233-58.
[http://dx.doi.org/10.1146/annurev.immunol.14.1.233] [PMID: 8717514]

[47] Pardoll DM. The blockade of immune checkpoints in cancer immunotherapy. Nat Rev Cancer 2012; 12(4): 252-64.
[http://dx.doi.org/10.1038/nrc3239] [PMID: 22437870]

[48] Hodi FS, O'Day SJ, McDermott DF, *et al.* Improved survival with ipilimumab in patients with metastatic melanoma. N Engl J Med 2010; 363(8): 711-23.
[http://dx.doi.org/10.1056/NEJMoa1003466] [PMID: 20525992]

[49] Keir ME, Butte MJ, Freeman GJ, Sharpe AH. PD-1 and its ligands in tolerance and immunity. Annu Rev Immunol 2008; 26: 677-704.
[http://dx.doi.org/10.1146/annurev.immunol.26.021607.090331] [PMID: 18173375]

[50] Freeman GJ, Long AJ, Iwai Y, *et al.* Engagement of the PD-1 immunoinhibitory receptor by a novel B7 family member leads to negative regulation of lymphocyte activation. J Exp Med 2000; 192(7):

1027-34.
[http://dx.doi.org/10.1084/jem.192.7.1027] [PMID: 11015443]

[51] Topalian SL, Drake CG, Pardoll DM. Targeting the PD-1/B7-H1(PD-L1) pathway to activate anti-tumor immunity. Curr Opin Immunol 2012; 24(2): 207-12.
[http://dx.doi.org/10.1016/j.coi.2011.12.009] [PMID: 22236695]

[52] Komura T, Sakai Y, Harada K, *et al.* Inflammatory features of pancreatic cancer highlighted by monocytes/macrophages and CD4+ T cells with clinical impact. Cancer Sci 2015; 106(6): 672-86.
[http://dx.doi.org/10.1111/cas.12663] [PMID: 25827621]

[53] Dong H, Strome SE, Salomao DR, *et al.* Tumor-associated B7-H1 promotes T-cell apoptosis: a potential mechanism of immune evasion. Nat Med 2002; 8(8): 793-800.
[http://dx.doi.org/10.1038/nm730] [PMID: 12091876]

[54] Butte MJ, Keir ME, Phamduy TB, Sharpe AH, Freeman GJ. Programmed death-1 ligand 1 interacts specifically with the B7-1 costimulatory molecule to inhibit T cell responses. Immunity 2007; 27(1): 111-22.
[http://dx.doi.org/10.1016/j.immuni.2007.05.016] [PMID: 17629517]

[55] Ralph C, Elkord E, Burt DJ, *et al.* Modulation of lymphocyte regulation for cancer therapy: a phase II trial of tremelimumab in advanced gastric and esophageal adenocarcinoma. Clin Cancer Res 2010; 16(5): 1662-72.
[http://dx.doi.org/10.1158/1078-0432.CCR-09-2870] [PMID: 20179239]

[56] Yamada Y, Nishina T, Iwasa S, *et al.* A phase I dose expansion trial of avelumab (MSB0010718C), an anti-PD-L1 antibody, in Japanese patients with advanced gastric cancer. J Clin Oncol 2015; 33(15_suppl): 4047-0.

[57] Nishina T, Shitara K, Iwasa S, *et al.* Safety, PD-L1 expression, and clinical activity of avelumab (MSB0010718C), an anti-PD-L1 antibody, in Japanese patients with advanced gastric or gastroesophageal junction cancer. J Clin Oncol 2016; 34(14_suppl): 168-0.

[58] Chung HC, Arkenau H-T, Wyrwicz L, *et al.* Avelumab (MSB0010718C; anti-PD-L1) in patients with advanced gastric or gastroesophageal junction cancer from JAVELIN solid tumor phase Ib trial: Analysis of safety and clinical activity J Clin Oncol 2016; 34(15_suppl): 4009-0.

[59] Muro K, Chung HC, Shankaran V, *et al.* Pembrolizumab for patients with PD-L1-positive advanced gastric cancer (KEYNOTE-012): a multicentre, open-label, phase 1b trial. Lancet Oncol 2016; 17(6): 717-26.
[http://dx.doi.org/10.1016/S1470-2045(16)00175-3] [PMID: 27157491]

[60] Fuchs CS, Ohtsu A, Tabernero J, *et al.* Preliminary safety data from KEYNOTE-059: Pembrolizumab plus 5-fluorouracil (5-FU) and cisplatin for first-line treatment of advanced gastric cancer. J Clin Oncol 2016; 34(15_suppl): 4037-0.

[61] Fuchs CS, Doi T, Jang RW-J, *et al.* KEYNOTE-059 cohort 1: Efficacy and safety of pembrolizumab (pembro) monotherapy in patients with previously treated advanced gastric cancer. J Clin Oncol 2017; 35(15_suppl): 4003-0.

[62] Catenacci DVT, Kim SS, Gold PJ, *et al.* A phase 1b/2, open label, dose-escalation study of margetuximab (M) in combination with pembrolizumab (P) in patients with relapsed/refractory advanced HER2+ gastroesophageal (GEJ) junction or gastric (G) cancer. J Clin Oncol 2017; 35(4_suppl): TPS219-.

[63] Chau I, Bendell JC, Calvo E, *et al.* Interim safety and clinical activity in patients (pts) with advanced gastric or gastroesophageal junction (G/GEJ) adenocarcinoma from a multicohort phase 1 study of ramucirumab (R) plus pembrolizumab (P). J Clin Oncol 2017; 35(4_suppl): 102-0.

[64] Le DT, Bendell JC, Calvo E, *et al.* Safety and activity of nivolumab monotherapy in advanced and metastatic (A/M) gastric or gastroesophageal junction cancer (GC/GEC): Results from the CheckMate-032 study. J Clin Oncol 2016; 34(4_suppl): 6-0.

[65]   Janjigian YY, Ott PA, Calvo E, *et al.* Nivolumab ± ipilimumab in pts with advanced (adv)/metastatic chemotherapy-refractory (CTx-R) gastric (G), esophageal (E), or gastroesophageal junction (GEJ) cancer: CheckMate 032 study. J Clin Oncol 2017; 35(15_suppl): 4014-0.

[66]   Janjigian YY, Adenis A, Aucoin J-S, *et al.* Checkmate 649: A randomized, multicenter, open-label, phase 3 study of nivolumab (Nivo) plus ipilimumab (Ipi) versus oxaliplatin plus fluoropyrimidine in patients (Pts) with previously untreated advanced or metastatic gastric (G) or gastroesophageal junction (GEJ) cancer. J Clin Oncol 2017; 35(4_suppl): TPS213-.

[67]   Kang Y-K, Satoh T, Ryu M-H, *et al.* Nivolumab (ONO-4538/BMS-936558) as salvage treatment after second or later-line chemotherapy for advanced gastric or gastro-esophageal junction cancer (AGC): A double-blinded, randomized, phase III trial. J Clin Oncol 2017; 35(4_suppl): 2-0.

[68]   Kloor M, Staffa L, Ahadova A, von Knebel Doeberitz M. Clinical significance of microsatellite instability in colorectal cancer. Langenbecks Arch Surg 2014; 399(1): 23-31.
[http://dx.doi.org/10.1007/s00423-013-1112-3] [PMID: 24048684]

[69]   Topalian SL, Hodi FS, Brahmer JR, *et al.* Safety, activity, and immune correlates of anti-PD-1 antibody in cancer. N Engl J Med 2012; 366(26): 2443-54.
[http://dx.doi.org/10.1056/NEJMoa1200690] [PMID: 22658127]

[70]   Brahmer JR, Tykodi SS, Chow LQ, *et al.* Safety and activity of anti-PD-L1 antibody in patients with advanced cancer. N Engl J Med 2012; 366(26): 2455-65.
[http://dx.doi.org/10.1056/NEJMoa1200694] [PMID: 22658128]

[71]   Lipson EJ, Sharfman WH, Drake CG, *et al.* Durable cancer regression off-treatment and effective reinduction therapy with an anti-PD-1 antibody. Clin Cancer Res 2013; 19(2): 462-8.
[http://dx.doi.org/10.1158/1078-0432.CCR-12-2625] [PMID: 23169436]

[72]   Le DT, Uram JN, Wang H, *et al.* PD-1 Blockade in Tumors with Mismatch-Repair Deficiency. N Engl J Med 2015; 372(26): 2509-20.
[http://dx.doi.org/10.1056/NEJMoa1500596] [PMID: 26028255]

[73]   Le DT, Durham JN, Smith KN, *et al.* Mismatch repair deficiency predicts response of solid tumors to PD-1 blockade. Science 2017; 357(6349): 409-13.
[http://dx.doi.org/10.1126/science.aan6733] [PMID: 28596308]

[74]   Le DT, Uram JN, Wang H, *et al.* Programmed death-1 blockade in mismatch repair deficient colorectal cancer. J Clin Oncol 2016; 34(15_suppl): 103-0.

[75]   Overman MJ, McDermott R, Leach JL, *et al.* Nivolumab in patients with metastatic DNA mismatch repair-deficient or microsatellite instability-high colorectal cancer (CheckMate 142): an open-label, multicentre, phase 2 study. Lancet Oncol 2017; 18(9): 1182-91.
[http://dx.doi.org/10.1016/S1470-2045(17)30422-9] [PMID: 28734759]

[76]   Andre T, Lonardi S, Wong KYM, *et al.* Combination of nivolumab (nivo) + ipilimumab (ipi) in the treatment of patients (pts) with deficient DNA mismatch repair (dMMR)/high microsatellite instability (MSI-H) metastatic colorectal cancer (mCRC): CheckMate 142 study. J Clin Oncol 2017; 35(15_suppl): 3531-0.

[77]   Bendell JC, Kim TW, Goh BC, *et al.* Clinical activity and safety of cobimetinib (cobi) and atezolizumab in colorectal cancer (CRC). J Clin Oncol 2016; 34(15_suppl): 3502-0.

[78]   Le DT, Uram JN, Wang H, *et al.* PD-1 blockade in mismatch repair deficient non-colorectal gastrointestinal cancers. J Clin Oncol 2016; 34(4_suppl): 195-0.

[79]   Cavalieri CC. Pembrolizumab in gastrointestinal (GI) malignancies with defective DNA mismatch repair (dMMR): A single institution experience. J Clin Oncol 2017; 35(15_suppl): e23081-.

[80]   Royal RE, Levy C, Turner K, *et al.* Phase 2 trial of single agent Ipilimumab (anti-CTLA-4) for locally advanced or metastatic pancreatic adenocarcinoma. J Immunother 2010; 33(8): 828-33.
[http://dx.doi.org/10.1097/CJI.0b013e3181eec14c] [PMID: 20842054]

[81]   Patnaik A, Kang SP, Rasco D, *et al.* Phase I Study of Pembrolizumab (MK-3475; Anti-PD-1 Monoclonal Antibody) in Patients with Advanced Solid Tumors. Clin Cancer Res 2015; 21(19): 4286-93.
[http://dx.doi.org/10.1158/1078-0432.CCR-14-2607] [PMID: 25977344]

[82]   van Elsas A, Hurwitz AA, Allison JP. Combination immunotherapy of B16 melanoma using anti-cytotoxic T lymphocyte-associated antigen 4 (CTLA-4) and granulocyte/macrophage colony-stimulating factor (GM-CSF)-producing vaccines induces rejection of subcutaneous and metastatic tumors accompanied by autoimmune depigmentation. J Exp Med 1999; 190(3): 355-66.
[http://dx.doi.org/10.1084/jem.190.3.355] [PMID: 10430624]

[83]   Hurwitz AA, Yu TF, Leach DR, Allison JP. CTLA-4 blockade synergizes with tumor-derived granulocyte-macrophage colony-stimulating factor for treatment of an experimental mammary carcinoma. Proc Natl Acad Sci USA 1998; 95(17): 10067-71.
[http://dx.doi.org/10.1073/pnas.95.17.10067] [PMID: 9707601]

[84]   Hurwitz AA, Foster BA, Kwon ED, *et al.* Combination immunotherapy of primary prostate cancer in a transgenic mouse model using CTLA-4 blockade. Cancer Res 2000; 60(9): 2444-8.
[PMID: 10811122]

[85]   Le DT, Lutz E, Uram JN, *et al.* Evaluation of ipilimumab in combination with allogeneic pancreatic tumor cells transfected with a GM-CSF gene in previously treated pancreatic cancer. J Immunother 2013; 36(7): 382-9.
[http://dx.doi.org/10.1097/CJI.0b013e31829fb7a2] [PMID: 23924790]

[86]   Aglietta M, Barone C, Sawyer MB, *et al.* A phase I dose escalation trial of tremelimumab (CP-675,206) in combination with gemcitabine in chemotherapy-naive patients with metastatic pancreatic cancer. Ann Oncol 2014; 25(9): 1750-5.
[http://dx.doi.org/10.1093/annonc/mdu205] [PMID: 24907635]

[87]   Kalyan A, Kircher SM, Mohindra NA, *et al.* Ipilimumab and gemcitabine for advanced pancreas cancer: A phase Ib study. J Clin Oncol 2016; 34(15_suppl): e15747-.

[88]   Borazanci EH, Hong DS, Gutierrez M, *et al.* Ibrutinib + durvalumab (MEDI4736) in patients (pts) with relapsed or refractory (R/R) pancreatic adenocarcinoma (PAC): A phase Ib/II multicenter study. J Clin Oncol 2016; 34(4_suppl): TPS484-.

[89]   Marks P, Rifkind RA, Richon VM, Breslow R, Miller T, Kelly WK. Histone deacetylases and cancer: causes and therapies. Nat Rev Cancer 2001; 1(3): 194-202.
[http://dx.doi.org/10.1038/35106079] [PMID: 11902574]

[90]   Haberland M, Montgomery RL, Olson EN. The many roles of histone deacetylases in development and physiology: implications for disease and therapy. Nat Rev Genet 2009; 10(1): 32-42.
[http://dx.doi.org/10.1038/nrg2485] [PMID: 19065135]

[91]   West AC, Johnstone RW. New and emerging HDAC inhibitors for cancer treatment. J Clin Invest 2014; 124(1): 30-9.
[http://dx.doi.org/10.1172/JCI69738] [PMID: 24382387]

[92]   Weichert W, Röske A, Gekeler V, *et al.* Histone deacetylases 1, 2 and 3 are highly expressed in prostate cancer and HDAC2 expression is associated with shorter PSA relapse time after radical prostatectomy. Br J Cancer 2008; 98(3): 604-10.
[http://dx.doi.org/10.1038/sj.bjc.6604199] [PMID: 18212746]

[93]   Weichert W, Röske A, Niesporek S, *et al.* Class I histone deacetylase expression has independent prognostic impact in human colorectal cancer: specific role of class I histone deacetylases in vitro and in vivo. Clin Cancer Res 2008; 14(6): 1669-77.
[http://dx.doi.org/10.1158/1078-0432.CCR-07-0990] [PMID: 18347167]

[94]   Weichert W, Röske A, Gekeler V, *et al.* Association of patterns of class I histone deacetylase expression with patient prognosis in gastric cancer: a retrospective analysis. Lancet Oncol 2008; 9(2):

139-48.
[http://dx.doi.org/10.1016/S1470-2045(08)70004-4] [PMID: 18207460]

[95]   Bolden JE, Peart MJ, Johnstone RW. Anticancer activities of histone deacetylase inhibitors. Nat Rev Drug Discov 2006; 5(9): 769-84.
[http://dx.doi.org/10.1038/nrd2133] [PMID: 16955068]

[96]   Falkenberg KJ, Johnstone RW. Histone deacetylases and their inhibitors in cancer, neurological diseases and immune disorders. Nat Rev Drug Discov 2014; 13(9): 673-91.
[http://dx.doi.org/10.1038/nrd4360] [PMID: 25131830]

[97]   Dokmanovic M, Marks PA. Prospects: histone deacetylase inhibitors. J Cell Biochem 2005; 96(2): 293-304.
[http://dx.doi.org/10.1002/jcb.20532] [PMID: 16088937]

[98]   Mottamal M, Zheng S, Huang TL, Wang G. Histone deacetylase inhibitors in clinical studies as templates for new anticancer agents. Molecules 2015; 20(3): 3898-941.
[http://dx.doi.org/10.3390/molecules20033898] [PMID: 25738536]

[99]   Nervi C, De Marinis E, Codacci-Pisanelli G. Epigenetic treatment of solid tumours: a review of clinical trials. Clin Epigenetics 2015; 7: 127.
[http://dx.doi.org/10.1186/s13148-015-0157-2] [PMID: 26692909]

[100] http://www.clinicaltrials.gov

[101] Doi T, Hamaguchi T, Shirao K, *et al.* Evaluation of safety, pharmacokinetics, and efficacy of vorinostat, a histone deacetylase inhibitor, in the treatment of gastrointestinal (GI) cancer in a phase I clinical trial. Int J Clin Oncol 2013; 18(1): 87-95.
[http://dx.doi.org/10.1007/s10147-011-0348-6] [PMID: 22234637]

[102] Fakih MG, Groman A, McMahon J, Wilding G, Muindi JR. A randomized phase II study of two doses of vorinostat in combination with 5-FU/LV in patients with refractory colorectal cancer. Cancer Chemother Pharmacol 2012; 69(3): 743-51.
[http://dx.doi.org/10.1007/s00280-011-1762-1] [PMID: 22020318]

[103] Wilson PM, El-Khoueiry A, Iqbal S, *et al.* A phase I/II trial of vorinostat in combination with 5-fluorouracil in patients with metastatic colorectal cancer who previously failed 5-FU-based chemotherapy. Cancer Chemother Pharmacol 2010; 65(5): 979-88.
[http://dx.doi.org/10.1007/s00280-009-1236-x] [PMID: 20062993]

[104] Yoo C, Ryu MH, Na YS, *et al.* Phase I and pharmacodynamic study of vorinostat combined with capecitabine and cisplatin as first-line chemotherapy in advanced gastric cancer. Invest New Drugs 2014; 32(2): 271-8.
[http://dx.doi.org/10.1007/s10637-013-9983-2] [PMID: 23712440]

[105] Yoo C, Ryu MH, Na YS, Ryoo BY, Lee CW, Kang YK. Vorinostat in combination with capecitabine plus cisplatin as a first-line chemotherapy for patients with metastatic or unresectable gastric cancer: phase II study and biomarker analysis. Br J Cancer 2016; 114(11): 1185-90.
[http://dx.doi.org/10.1038/bjc.2016.125] [PMID: 27172248]

[106] Okada K, Hakata S, Terashima J, Gamou T, Habano W, Ozawa S. Combination of the histone deacetylase inhibitor depsipeptide and 5-fluorouracil upregulates major histocompatibility complex class II and p21 genes and activates caspase-3/7 in human colon cancer HCT-116 cells. Oncol Rep 2016; 36(4): 1875-85.
[http://dx.doi.org/10.3892/or.2016.5008] [PMID: 27509880]

[107] Moffat D, Patel S, Day F, *et al.* Discovery of 2-(6-{[(6-fluoroquinolin-2-yl)methyl]amino} bicyclo[3.1.0]hex-3-yl)-N-hydroxypyrim idine-5-carboxamide (CHR-3996), a class I selective orally active histone deacetylase inhibitor. J Med Chem 2010; 53(24): 8663-78.
[http://dx.doi.org/10.1021/jm101177s] [PMID: 21080647]

[108] Banerji U, van Doorn L, Papadatos-Pastos D, *et al.* A phase I pharmacokinetic and pharmacodynamic

study of CHR-3996, an oral class I selective histone deacetylase inhibitor in refractory solid tumors. Clin Cancer Res 2012; 18(9): 2687-94.
[http://dx.doi.org/10.1158/1078-0432.CCR-11-3165] [PMID: 22553374]

[109]   A Phase I/II study to evaluate safety, tolerability, pharmacokinetics and efficacy of Resminostat (4SC-201) in combination with a second-line treatment in patients with K-ras mutated advanced colorectal carcinoma Available from: https://clinicaltrialsgov/show/NCT01277406

[110]   Iwahashi S, Ishibashi H, Utsunomiya T, *et al.* Effect of histone deacetylase inhibitor in combination with 5-fluorouracil on pancreas cancer and cholangiocarcinoma cell lines. J Med Invest 2011; 58(1-2): 106-9.
[http://dx.doi.org/10.2152/jmi.58.106] [PMID: 21372494]

[111]   Iwahashi S, Utsunomiya T, Imura S, *et al.* Effects of valproic acid in combination with S-1 on advanced pancreatobiliary tract cancers: clinical study phases I/II. Anticancer Res 2014; 34(9): 5187-91.
[PMID: 25202113]

[112]   Richards DA, Boehm KA, Waterhouse DM, *et al.* Gemcitabine plus CI-994 offers no advantage over gemcitabine alone in the treatment of patients with advanced pancreatic cancer: results of a phase II randomized, double-blind, placebo-controlled, multicenter study. Ann Oncol 2006; 17(7): 1096-102.
[http://dx.doi.org/10.1093/annonc/mdl081] [PMID: 16641168]

[113]   Lee HS, Park SB, Kim SA, *et al.* A novel HDAC inhibitor, CG200745, inhibits pancreatic cancer cell growth and overcomes gemcitabine resistance. Sci Rep 2017; 7: 41615.
[http://dx.doi.org/10.1038/srep41615] [PMID: 28134290]

[114]   Fotheringham S, Epping MT, Stimson L, *et al.* Genome-wide loss-of-function screen reveals an important role for the proteasome in HDAC inhibitor-induced apoptosis. Cancer Cell 2009; 15(1): 57-66.
[http://dx.doi.org/10.1016/j.ccr.2008.12.001] [PMID: 19111881]

[115]   Khan O, Fotheringham S, Wood V, *et al.* HR23B is a biomarker for tumor sensitivity to HDAC inhibitor-based therapy. Proc Natl Acad Sci USA 2010; 107(14): 6532-7.
[http://dx.doi.org/10.1073/pnas.0913912107] [PMID: 20308564]

[116]   Piekarz RL, Frye R, Turner M, *et al.* Phase II multi-institutional trial of the histone deacetylase inhibitor romidepsin as monotherapy for patients with cutaneous T-cell lymphoma. J Clin Oncol 2009; 27(32): 5410-7.
[http://dx.doi.org/10.1200/JCO.2008.21.6150] [PMID: 19826128]

[117]   Milde T, Lodrini M, Savelyeva L, *et al.* HD-MB03 is a novel Group 3 medulloblastoma model demonstrating sensitivity to histone deacetylase inhibitor treatment. J Neurooncol 2012; 110(3): 335-48.
[http://dx.doi.org/10.1007/s11060-012-0978-1] [PMID: 23054560]

[118]   Yeo W, Chung HC, Chan SL, *et al.* Epigenetic therapy using belinostat for patients with unresectable hepatocellular carcinoma: a multicenter phase I/II study with biomarker and pharmacokinetic analysis of tumors from patients in the Mayo Phase II Consortium and the Cancer Therapeutics Research Group. J Clin Oncol 2012; 30(27): 3361-7.
[http://dx.doi.org/10.1200/JCO.2011.41.2395] [PMID: 22915658]

[119]   Zheng H, Zhao W, Yan C, *et al.* HDAC Inhibitors Enhance T-Cell Chemokine Expression and Augment Response to PD-1 Immunotherapy in Lung Adenocarcinoma. Clin Cancer Res 2016; 22(16): 4119-32.
[http://dx.doi.org/10.1158/1078-0432.CCR-15-2584] [PMID: 26964571]

[120]   Hopewell EL, Zhao W, Fulp WJ, *et al.* Lung tumor NF-κB signaling promotes T cell-mediated immune surveillance. J Clin Invest 2013; 123(6): 2509-22.
[http://dx.doi.org/10.1172/JCI67250] [PMID: 23635779]

[121]   Magner WJ, Kazim AL, Stewart C, *et al.* Activation of MHC class I, II, and CD40 gene expression by

histone deacetylase inhibitors. J Immunol 2000; 165(12): 7017-24.
[http://dx.doi.org/10.4049/jimmunol.165.12.7017] [PMID: 11120829]

[122] Maeda T, Towatari M, Kosugi H, Saito H. Up-regulation of costimulatory/adhesion molecules by histone deacetylase inhibitors in acute myeloid leukemia cells. Blood 2000; 96(12): 3847-56.
[PMID: 11090069]

[123] Khan AN, Gregorie CJ, Tomasi TB. Histone deacetylase inhibitors induce TAP, LMP, Tapasin genes and MHC class I antigen presentation by melanoma cells. Cancer Immunol Immunother 2008; 57(5): 647-54.
[http://dx.doi.org/10.1007/s00262-007-0402-4] [PMID: 18046553]

[124] Setiadi AF, Omilusik K, David MD, *et al.* Epigenetic enhancement of antigen processing and presentation promotes immune recognition of tumors. Cancer Res 2008; 68(23): 9601-7.
[http://dx.doi.org/10.1158/0008-5472.CAN-07-5270] [PMID: 19047136]

[125] Skov S, Pedersen MT, Andresen L, Straten PT, Woetmann A, Odum N. Cancer cells become susceptible to natural killer cell killing after exposure to histone deacetylase inhibitors due to glycogen synthase kinase-3-dependent expression of MHC class I-related chain A and B. Cancer Res 2005; 65(23): 11136-45.
[http://dx.doi.org/10.1158/0008-5472.CAN-05-0599] [PMID: 16322264]

[126] Armeanu S, Bitzer M, Lauer UM, *et al.* Natural killer cell-mediated lysis of hepatoma cells via specific induction of NKG2D ligands by the histone deacetylase inhibitor sodium valproate. Cancer Res 2005; 65(14): 6321-9.
[http://dx.doi.org/10.1158/0008-5472.CAN-04-4252] [PMID: 16024634]

[127] Beg AA, Gray JE. HDAC inhibitors with PD-1 blockade: a promising strategy for treatment of multiple cancer types? Epigenomics 2016; 8(8): 1015-7.
[http://dx.doi.org/10.2217/epi-2016-0066] [PMID: 27410519]

[128] Ma T, Galimberti F, Erkmen CP, *et al.* Comparing histone deacetylase inhibitor responses in genetically engineered mouse lung cancer models and a window of opportunity trial in patients with lung cancer. Mol Cancer Ther 2013; 12(8): 1545-55.
[http://dx.doi.org/10.1158/1535-7163.MCT-12-0933] [PMID: 23686769]

[129] Kroemer G, Galluzzi L, Zitvogel L, Fridman WH. Colorectal cancer: the first neoplasia found to be under immunosurveillance and the last one to respond to immunotherapy? OncoImmunology 2015; 4(7): e1058597.
[http://dx.doi.org/10.1080/2162402X.2015.1058597] [PMID: 26140250]

CHAPTER 4

Long Non-Coding RNAs in Cancer Progression: Implication for Anti-Cancer Therapy

Chang Gong[1,*], **Zihao Liu**[1], **Andrew J. Sanders**[2] and **Wen G. Jiang**[2,*]

[1] *Guangdong Provincial Key Laboratory of Malignant Tumor Epigenetic and Gene Regulation, Breast Tumor Center, Sun Yat-sen Memorial Hospital, Sun Yat-sen University, Guangzhou, China*

[2] *Cardiff China Medical Research Collaborative, Cardiff University School of Medicine, Cardiff University, Heath Park, Cardiff, UK*

Abstract: long non-coding RNAs (lncRNAs) are a diverse group of functional RNAs which were once regarded as "transcription noise". However, the latest evidence has revealed that lncRNAs are often cancer-specific and the expression of lncRNAs is frequently aberrant in a variety of cancers. Several studies have highlighted the fundamental roles of lncRNAs in cancer proliferation, self-renewal, chemotherapy resistance and metastasis. Thus, lncRNAs lie at the crossroad of tumorigenesis. Recently, several articles have explored the molecular mechanisms notably governing lncRNAs in multiple biological processes of the tumor. LncRNAs in the nucleus act in *trans* or in *cis* to promote the transcription of tumorigentic genes. LncRNAs in the nucleus play an important role in alternative splicing or genetic imprinting. In addition, lncRNAs residing in the cytoplasm can exert their peculiar regulation on mRNA, microRNA or protein via post-transcriptional modification. Targeting lncRNAs can reverse cancer progression and might be promising for anti-cancer treatment. Several different approaches to target lncRNAs including: small interfering RNAs, antisense oligonucleotide, aptamer, small molecular inhibitors and CRISPR systems have been considered for therapeutic purposes. This review will focus on the involvement of lncRNAs in cancer progression and provide valuable therapeutic information to target lncRNAs for anti-cancer treatment.

Keywords: Llong Non-Coding RNA, Epigenetics, LncRNA Classification, LncRNA Location, LncRNA Structure, Conformation Structure, Transcriptional Regulation, *in cis* LncRNA, *in trans* LncRNA, Aalternative Splicing, Genetic Imprinting, Ppost-Transcriptional Regulation, RNA Decay, Endogenous Competing RNA, Cancer, Proliferation, Stemness, Chemotherapy Resistance, Migration, Invasion, Metastasis, Anti-Cancer Ttherapy, RNA Interference,

* **Corresponding author Chang Gong:** Sun Yat-sen Memorial Hospital, Sun Yat-sen University, 107 Yanjiang West Road, Guangzhou, 510120, P.R. China; E-mails: changgong282@163.com or gchang@mail.sysu.edu.cn
* **Corresponding author Wen G. Jiang:** Cardiff China Medical Research Collaborative, Cardiff University School of Medicine, Cardiff University, Heath Park, Cardiff, CF14 4XN, UK; E-mails: jiangw@cardiff.ac.uk

Atta-ur-Rahman (Ed.)

Antisense Oligonucleotides, Apatmers, Drug Delivery, Small Molecule, CRISPR System.

INTRODUCTION

In recent years, with the development of next-generation sequencing methods and the explosion of large-scale sequencing projects such as: Encyclopedia of DNA Elements (ENCODE), Functional Annotation of the Mammalian genome project (FANTOM), The Cancer Genome Atlas (TCGA) and so on, we have made great discoveries on the complexities of our genome. Our understanding of our genome has dramatically changed, we know that only a small portion of genes, occupying nearly 2-3% of the genome, are protein coding genes [1]. The most exciting outcome of this advance data was that the high reads of transcriptome sequencing can be aligned to the non-coding regions which are highly conserved from worms to human [2, 3]. This indicates that transcripts from these regions are potentially functional [4]. Indeed, some of these non-coding RNAs appear to exert their regulator roles in human cells [5]. Non-coding RNAs are termed as RNA transcripts that cannot encode proteins. In the past, non-coding RNAs were always referred as RNAs of translation machinery such as: transfer RNAs and ribosomal RNAs. Recently, the content of non-coding RNAs extends from translation components to long non-coding RNAs (lncRNAs).

Long non-coding RNAs are defined as non-coding RNAs containing over 200 nucleic acids and this concept was firstly proposed in 2002 [6]. Due to the definition of lncRNAs, they can diversify from 2kb long such as NF-KappaB Interacting LncRNA (NKILA) to even more than 90kb long such as Kcnq1ot1. Although lncRNAs do not code proteins, they do exert their extraordinary regulatory role in gene transcriptional regulation, chromatin remodeling, genetic imprinting and post-transcriptional regulation [7]. Since their mechanisms of action are very flexible, they can participate in different physiological processes, such as embryo development, endocrine system function and aging [8, 9]. They also play a fundamental role in several diseases such as: cardiovascular disease, central nerve system disorder and metabolic disease and cancers [9, 10]. Targeting lncRNAs might be promising for anti-cancer treatment. Several approaches which can reverse abnormal expression of lncRNAs such as: small interfering RNAs, antisense oligonucleotide, aptamer, CRISPR/Cas9 system or approaches aborting molecule to molecule interactions have been tested in preclinical models or clinical trials. This chapter focuses on the involvement of lncRNAs in cancer and provides valuable therapeutic information to target lncRNAs for anti-cancer treatment.

OVERVIEW OF LONG NON-CODING RNA BIOLOGY

Due to the lack of an accurate definition of lncRNAs, lncRNAs are arbitrarily defined as transcripts longer than 200nt in size to distinguish them from small non-coding RNAs such as: transfers RNA (tRNA), microRNA (miRNA), piwi-interacting RNA (piRNA) and small nucleolar RNAs (snoRNAs) [3]. This definition has been widely accepted worldwide. However, there are some flaws in this definition. Recently, research has revealed that protein coding regions reside in lncRNA and can encode functional micropeptides [11].

LncRNAs enjoy a high abundance as well as heterogeneity and, to some extent, reflect the complexity of their structure. An effective classification to categorize these heterogeneous and diverse lncRNAs is still lacking. Based on the length of lncRNAs, they can be categorized into three groups: long intergenic non-coding RNAs, very long intergenic non-coding RNAs and macro long non-coding RNAs. According to the relationship between lncRNAs and their parent genes or neighbor genes, lncRNAs can be classified into five groups as: intergenic long non-coding RNAs, antisense long non-coding RNAs, bidirectional long non-coding RNAs, intronic long non-coding RNAs and sense long non-coding RNAs. In addition to this, lncRNAs can also be divided into groups based on their action of mechanism as: enhancer long non-coding RNAs which function through enhancing gene expression; competing endogenous long non-coding RNAs that sponge miRNAs and regulate gene expression post-transcriptionally; miRNA/piRNA primary transcripts long non-coding RNAs which can be spliced into miRNA or piRNA and *et al* [7, 12, 13]. However, these classification methods mentioned above have their own limitations. For instance, lncRNA AIRN which is transcribed from insulin-like growth factor 2 receptor (IGF2R) belongs to antisense lncRNA, on the other hand, it belongs to macro lncRNA based on its size. AIRN interacts with its sense DNA and act *in cis* via recruiting histone regulators. LncRNA ANRIL belongs to intergenic lncRNA, but ANRIL not only regulates the expression of adjacent genes *in cis*, but also promotes the expression of distant genes *in trans* [14].

In contrast to message RNAs (mRNAs), most lncRNAs tend to stay in the nucleus rather than translocating to the cytoplasm [15]. It should be noted that the peculiar sequences of lncRNAs and their secondary or tertiary structures might attribute to their nucleus retention. A triple helix can form between the U-rich internal loop and the poly A tails of lncRNA metastasis-associated lung adenocarcinoma transcript 1 (MALAT1) [16]. Mutation of a novel RNA motif of lncRNA BMP2-OP1-responsive gene (BORG) to a scrambled sequence totally abrogates the nuclear localization of this lncRNA [17]. In addition, the interactions between lncRNAs and specific proteins also trap lncRNAs to retain them in the nucleus. Heterogeneous nucleus ribonucleoproteins U (hnRNP U) can be recruited by lncRNA X-inactive specific transcript (Xist) to the inactive X chromosome. HnRNP U knockdown will result in the failure of inactive X chromosome and the diffusion of Xist into nucleoplasm from inactive X chromosome [18]. This evidence suggests that hnRNP U is required for localization. LncRNA Firre resides in the *trans*-chromosomal loci and regulates pluripotency gene expression *in trans* through recruiting hnRNP U. Knockdown of HnRNP U will lead to the loss of co-localization of Firre and *trans*-chromosomal loci and translocation of Firre into the cytoplasm [19]. Thus such proteins play an important role in lncRNA location.

One of the major features of lncRNA is that they can be folded into RNA molecules with stable secondary and tertiary structures. Hydrogen bonds can be formed on the different faces of single stranded RNA such as: Watson-Crick face, ribose face and Hoogsteen face [20]. Although these interactions are rather weaker than chemical bonds, they collectively contribute to the formation of secondary structures of lncRNAs. Although most conserved secondary structures have been characterized in rRNAs and tRNAs, lncRNAs also turn out to have secondary structures such as hairpin structure, helix loop, bulges and so on [21]. Much attention has been drawn to the higher-order structures of lncRNA since these RNA structures dictate their fundamental function. For example, the secondary structure of lncRNA maternally expressed gene 3 (MEG3) is necessary for its biological function [22]. The steroid receptor RNA activator (SRA) lncRNA, has four RNA domains which consist of several secondary structures. It coactivates several hormone receptors through its functional domains and promotes breast cancer tumorigenesis [23].

MECHANISM OF ACTION

Assembling to proteins, the mechanisms of action of lncRNAs are highly dependent on their subcellular location. Some lncRNAs must be translocated into the cytoplasm and function as post-transcriptional regulators of proteins, regulate mRNA translation, act as miRNA/piRNA primary transcripts or form competing endogenous RNAs to sponge miRNAs. Many lncRNAs prefer to reside in the nucleus. Nuclear lncRNAs can regulate gene expression *in cis* or *in trans*, impact on the splicing machinery of mRNA and affect the integrity of subnuclear organelles such as paraspeckles [7, 12]. The models for the mechanisms of action of lncRNAs are presented in Fig. (**1**).

Nucleic lncRNAs Act *in cis*

Of note, one group of lncRNAs arising from gene promoters will not detach from their transcriptional sites, but accumulate at the site and epigenetically regulate gene expression (*in cis*). These lncRNAs play a fundamental role in regulating the expression of their adjacent genes. They can bind to the promoters of adjacent genes and recruit transcriptional factors, histone modifiers and chromosome organizers to control gene expression.

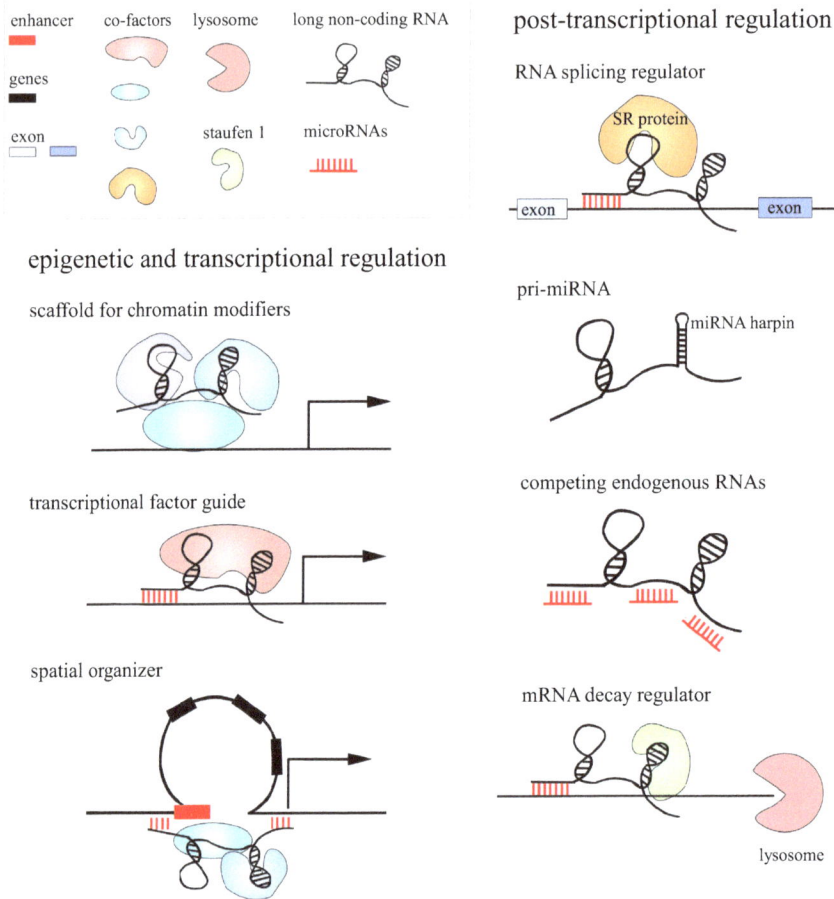

Fig. (1). Models for mechanism of action of lncRNAs. LncRNAs can exert their regulatory roles through transcription, post-transcription, translation or post-translational levels. For instance, lncRNA can be processed into miRNA/piRNA, form competing endogenous RNAs to sponge miRNAs, regulate alternative splicing or affect mRNA decay. LncRNAs can regulate gene expression *in cis* or *in trans* through acting as scaffold, transcriptional factor guide or become spatial organizers.

Enhancers were first characterized as DNA elements which locate far away from the transcriptional start sites of protein coding genes. They can bind with transcriptional factors and are required for positive regulation of gene expression in a spatial manner via unique recognition sequences [24]. Recent findings suggest that lncRNAs regulate focal gene expression which is much more similar to enhancers *in cis*. CREB-binding protein can interact with enhancers and recruit polymerase II to actively transcribe a class of bidirectional enhancer lncRNAs with enhancer motif. The positive correlation between the expression level of enhancer lncRNAs and nearby mRNA synthesis was also observed [25]. Knockdown of enhancer lncRNAs results in the reduced expression of their nearby protein coding genes. This indicates that enhancer lncRNAs are required for adjacent gene expression and they potentially function *in cis*. Recently, insights into the mechanism of action of *in cis* lncRNAs were provided. LncRNA HOXA transcript at the distal tip (HOTTIP) which is transcribed from 5' region of *HOXA* cluster can bind to HOXA by forming a chromosomal loop. By recruiting histone modifier WD Repeat Domain 5 (WDR5) and mixed lineage leukemia (MLL) protein across *HOXA* gene locus, HOTTIP promotes histone H3 lysine 4 trimethylation and drives HOX gene expression [26]. LncRNA LUNAR1 (leukemia-induced non-coding activator RNA) is proximate to insulin-like growth factor receptor 1 (IGF1R) and pyroglutamyl peptidase 1-like (PGPEP1L) protein, but LUNAR1 only binds to the promoter of IGF1R by forming a loop crossing whole *PGPEP1L* locus spatially. Via recruiting NOTCH transcriptional complex, polymerase II and co-factors, LUNAR1 contributes to sustainable expression of IGF1R [27]. LncRNA Colorectal Cancer Associated Transcript 1 (CCAT1), which is transcribed by super enhancer, can regulate downstream gene MYC *in cis* through recruiting chromatin modifier CCCTC binding factor (CTCF) [28]. The phenomenon that enhancer lncRNAs can promote gene expression *in cis* is quite universal in our genome. Besides, these enhancer lncRNAs are necessary for local gene expression. Knockdown of CCAT1 can inhibit the interaction between MYC enhancers and its promoter [28]. Chromosome conformation capture technique also confirms interaction between enhancer lncRNAs ncRNA-activating (ncRNA-a) and their target gene promoters [29].

Natural antisense transcripts (NATs) of lncRNAs are other classes of *cis*-acting modulators. Since 50-70% of general sense sequences have antisense transcripts, the scale of antisense transcription might be larger than we previously thought [3, 30]. In addition, more than half of natural antisense transcripts do not code proteins. Antisense lncRNAs indeed control gene expression at different levels such as: transcriptional regulation, translational regulation, RNA editing and *et al*. Antisense lncRNAs in the nucleus can anchor themselves to the DNA via their complementary sequences and control gene expression *in cis*. Antisense intronic non-coding RASSF1 (ANRASSF1) which is transcribed from the antisense strand of RAS-association domain family member 1A (RASSF1A) binds to the RASSF1A promoter specifically through forming DNA/RNA triple helices. Polycomb repressive complex 2 (PRC2), an epigenetic regulator, interacts with this lncRNA and represses RASSF1A [31]. CDKN2B (p15^{INK4b})-p14^{ARF}-CDKN2A (p16^{INK4a}) cluster encodes three tumor suppressor proteins: p14, p15 and p16 whose expression are restricted to PRC family and histone H3 lysine 27 methylation (H3K27me) [32]. LncRNA antisense non-coding RNA in the INK4 locus (ANRIL) encoded by *CDKN2B (p15^{INK4b})* antisense transcript 1 is found to interact with chromobox 7 (CBX7), a counterpart of PRC. Depletion of ANRIL impairs the function of CBX7 which inhibits *CDKN2B-CDKN2A* expression [33]. Similarly, lncRNA antisense of insulin-like growth factor-2 receptor RNA non-coding (AIRN) controls the level of its parental gene in *cis* via interacting with IGF2R promoter [34].

Nucleic lncRNAs Act *in trans*

In addition to acting *in cis*, lncRNAs also relocate globally in the nucleus and bind to their target DNA sequence far away from their synthesis sites. *HOX* genes contain four clusters *HOXA*, *HOXB*, *HOXC* and *HOXD* which locate at chromosome 7, 17, 12 and 2 individually. The highly conserved HOX proteins play an important role in embryo development as well as cellular differentiation [35]. Unlike HOTTIP encoded by the *HOXA* gene cluster, lncRNA HOX antisense intergenic RNA (HOTAIR) emerges as *trans*-acting lncRNA to regulate HOXD expression. HOTAIR is an antisense lncRNA. The mature form of HOTAIR produces higher-order structures at both ends individually and acts as a scaffold for histone modifiers. The 3' domain of HOTAIR binds with corepressor REST which is responsible for demethylation of histone H3 lysine 4, while the 5' domain interacts with PRC2 that takes charge of histone H3 lysine 27 methylation. HOTAIR can guide these histone modifiers to *HOXD* locus and repress the expression of HOXD, but has no impact on the *HOXC* gene cluster [36]. However, HOTAIR can occupy sites beyond the HOXD locus with more than 800 identified binding sites. Similarly, long intergenic non-coding RNAs p21 (lincRNA p21), which is induced by p53, can transcriptionally represses hundreds of its target genes *in trans* via recruiting heterogeneous nucleus ribonucleoprotein K (hnRNPK) [37].

Nucleic lncRNAs Function in Alternative Splicing

Apart from acting in *trans* or *in cis*, nuclear lncRNAs also function through other mechanisms. Early evidence suggesting that lncRNAs may regulate alternative splicing of premessenger RNA (pre-mRNA) was first proposed by NATs. In rats as well as humans, thyroid hormone receptors are encoded by two loci, TRα and TRβ genes for humans, Thra and Thrb for rats, alias NR1A1 and NR1A2 or also termed as: c-erbAα and c-erbAβ. C-erbAα locus can transcribe two genes erbAα1 and erbAα2 which are distinguishable through alteration splicing of exon 3 [38]. Interestingly, lncRNA Rev-ErbAα, transcribed from the antisense strand of erbAα2, is very complementary to α2 mRNA but not to α1. By base paring with exon 3, Rev-ErbAα blocks the alternative splicing of α2 and forms the non-overlapping α1 transcripts [39]. To date, other lncRNAs are also found to control alternative splicing processes. The serine/arginine (SR) splicing factors can be recruited to the transcriptional sties of mRNA and regulate the alternative splicing processes. Since the splicing machinery is tightly regulated, a slight change of the SR proteins level or the phosphorylation status can reset the balance. MALAT1 associates with different classes of SR proteins and regulates the phosphorylation status of SR. Depletion of MALAT1 leads to the displacement of splicing components in nucleus speckles and alters the pattern of pre-mRNA splicing [40]. In addition to regulating alternative splicing, suborganelle integrity is dependent on lncRNAs. LncRNA nucleus-enriched autosomal transcript 1 (NEAT1) is a highly abundant and conserved lncRNA which always colocalizes with nuclear paraspeckles [41, 42]. By acting as a protein scaffold, several proteins such as SFPQ, FUS and RBM dock to NEAT1 and form integrated paraspeckles [43]. NEAT1 is believed to be a necessary structural backbone for paraspeckles since paraspeckles fail to assemble in NEAT1 knockout mice.

Cytoplasmic lncRNAs Regulate RNA Sstability

Some lncRNAs have to be transported to the cytoplasm to execute their regulatory roles. In the cytoplasm, a set of lncRNAs regulates the half life of mRNA that is important in mRNA metabolism. The first evidence that lncRNA regulates mRNA decay was reported in NATs. The 5' as well as 3' untranslated region (UTR) of hypoxia induced factor 1α genes can be spliced into two antisense lncRNA transcripts; 5' (5'aHIF-1α) and 3' (3'aHIF-1α) under different kinds of stress. 5'aHIF-1α transcript has a 5' cap and a poly A tail while 3'aHIF-1α lacks a 5' cap or a poly A tail. The two antisense transcripts of HIF-1α show totally different regulatory mechanisms. Of note, high level of 3'aHIF-1α is negatively correlated with HIF1α expression and predicts poor prognosis in breast cancer. This indicates the potential involvement of 3'aHIF-1α in post-transcriptional regulation of HIF1α mRNA. After hybridization to 3'aHIF-1α, the hairpin structure of the AU rich element within the 3' UTR of HIF1α mRNA will be disrupted, and in turn, allows the AU rich element binding proteins to interact with the single strand of HIF1α mRNA [44]. These AU rich element binding proteins trigger proteasome-dependent mRNA degradation [45]. Staufen 1 (STAU1) is a double-stranded RNA binding protein and can degrade active mRNA via STAU1 binding sites in the 3' UTR of mRNA. Interestingly, cytoplasmic lncRNAs half-STAU--binding site RNAs (1/2-sbsRNAs) interact with mRNA specifically, form double strand STAU1 binding elements, and, in turn, conduct STAU1-mediated mRNA decay [46]. In addition to the Alu elements, mRNA and lncRNA interactions also happen via other elements. LncRNA terminal differentiation-induced ncRNA (TINCR) interacts with its target mRNA through a 25-nucleotide TINCR motif and recruits STAU1 to decay mRNA. The TINCR motif is strongly required for TINCR-mRNA interaction [47].

Cytoplasmic lncRNA in Protein Modification

Cytoplasmic lncRNAs also play a fundamental role in regulating post-translational events of proteins, such as: phosphorylation and ubiquitination. For instance, unactivated NF-κB-p65/NF-κB-p50 heterodimers primarily interact with the inhibitory factor IκB. IkB kinases (IKK) can phosphorylate IκB which subsequently activates NF-κB pathway in response to rapid stimuli [48]. An lncRNA NF-κB Interacting LncRNA (NKILA) is induced by inflammatory stimuli. NKILA interacts with unactivated NF-κB complex through two harpin structures [49]. Because NKILA masks the phosphorylation site of IκB, IκB cannot be phosphorylated by IKK and in turn contributes to NF-κB pathway inhibition [49]. LncRNA lnc-DC is specifically expressed by dendritic cells. In the cytoplasm, lnc-DC associates with signal transducer and activator of transcription 3 (STAT3) via its 3' stem loop and prevents STAT3 dephophorylation from SH2 domain-containing protein tyrosine phosphatase-1 (SHP1). The phosphorylation of STAT3 is important for translocation of STAT3 and maintaining STAT3 pathway activation [50]. Similarly, lincRNA p21 also exerts its regulatory roles in post-translational protein modifications. Von Hippel-Lindau (VHL) protein recognizes HIF-1α which subsequently degrades HIF-1α *via* the ubiquitination-dependent proteasome pathway. LincRNA p21 docks into the binding domain of HIF-1α as well as VHL protein. This disrupts VHL-HIF-1α interaction and abrogates VHL associated ubiquitinating degradation which causes a positive feedback loop between HIF-1α and LincRNA p21 [51].

Competing Endogenous RNAs

Identical miRNAs can bind to lncRNAs through the imperfect complementary base pairing with lncRNA sequences. These lncRNAs compete with miRNA binding and "sponge" these miRNAs to avoid miRNA-dependent degradation of mRNAs. These lncRNAs are so called competing endogenous RNAs (ceRNAs). For instance, the lncRNA highly up-regulated in liver cancer (HULC) sponges miR-372 and promotes the expression of downstream mRNAs of miR-372 [52]. LncRNA MD1 sponges miR-133 and miR-135. It governs the expression of Mastermind-like protein 1 (MAML1) and MADS box transcription enhancer factor 2 (MEF2C) which are downstream targets of miR-133 and miR-135 [53]. Other examples of long non-coding ceRNAs have also been reported [54 - 56].

Endogenous Small Interfering RNA Formation

A genome-wide scale analysis predicts that there are approximately 100 lncRNAs that potentially encode endogenous small interfering RNA (siRNA) [57]. Because harpin structures may form between the highly complementary regions within lncRNAs and can be processed by RNA-induced silencing complex (RISC) or

RNA binding proteins to become endogenous siRNAs. Indeed, exon 1 of lncRNA H19 can be processed into two conserved miRNAs: miR-675-3p and miR-675-5p and H19 is a functional turnover by being spliced into the mature miR-675 [58]. LncRNA MD1 can also give rise to miR-133 precursor. Hu antigen R (HuR) is a RNA-binding protein. HuR is recruited to pri-miR-133 stem loop structure of lncRNA MD1 and splices lncRNA MD1 into pre-miR-133b [59].

LONG NON-CODING RNA IN CANCER

In a molecular biology perspective, cancers are now regarded as highly heterogeneous diseases involving genome instability, epigenetic alternations, dysregulation of multiple pathways and metabolic reprogramming which often govern the transformation of cancer cellular processes such as: self-renewal, resisting cell death, proliferation, angiogenesis, chemotherapy resistance and metastasis [60]. Since the aberrant expression of lncRNAs is found to be associated with genome stability and can regulate oncogenes or tumor suppressors at several levels including the transcriptional, post-transcriptional, post-translational level and *et al*, lncRNAs lie at the crossroad of cancer progression (Fig. **2**).

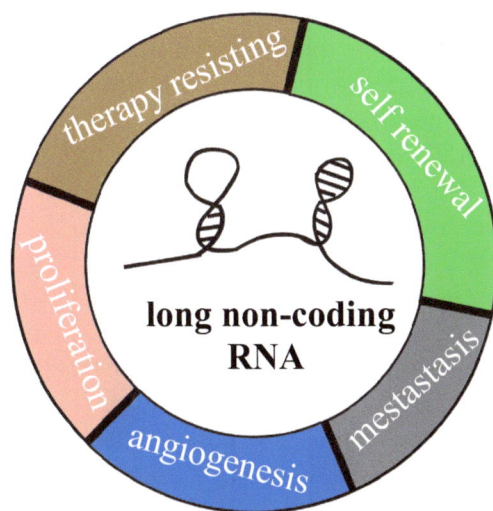

Fig. (2). LncRNAs lie at the crossroad of cancer progression. LncRNAs play a fundamental role in cancer self-renewal, proliferation, angiogenesis, chemotherapy resistance and metastasis through different molecular mechanisms.

We start to recognize that the properties of cancer initiating cells may resemble embryonic stem cells. Several key lncRNAs which function in embryo stem cells are also under investigation in cancers. Depletion of a highly conserved vertebral lncRNA, megamind, in Zebrafish results in the defects of brain and eyes [61]. Ablation of its ortholog in mouse, lncRNA TUNA, results in neural differentiation and stem cell marker down-regulation [62]. These suggest this lncRNA plays a vital role in embryo stem cell maintenance. Interestingly, its human transcript linc00617 is up-regulated in breast cancer cells. Up-regulation of this lncRNA enriches $CD44^{high}/CD24^{low}$ breast cancer stem cells and can enhance tumor sphere formation [63]. Similarly, Gata6 long noncoding RNA (lncGata6) is highly expressed by lgr5 positive intestinal stem cells (ISCs) and maintains stemness of ISCs. When expressed by cancer stem cells of colorectal cancer, lncGata6 promotes colorectal cancer initiation and progression [64]. Estradiol-induced H19 induces cancer stem-like traits of papillary thyroid cancer cells and high expression of H19 predicts poor overall survival of papillary thyroid cancer patients [65]. Several lncRNAs are found to maintain cancer cell "stemness" [63, 66]. LncRNAs also confer growth advantages to cancer cells by regulating cancer cell proliferation and angiogenesis. For instance, lncRNA prostate cancer–associated ncRNA transcript 1 (PCAT-1), which appears to display prostate cancer-specific expression patterns, regulates cancer proliferation by recruiting PRC2 [67]. LncRNA SAMMSON promotes melanoma cell survival by activating microphthalmia-associated transcription factor (MITF) *in cis* [68]. Knockdown of SAMMSON will inhibit melanoma growth *in vitro* and *in vivo*. MALAT1 promotes the proliferation of several cancer types including: multiple myeloma, hepatocellular carcinoma, thyroid cancer and colorectal cancer [69 - 72]. Silencing of MALAT1 alters the phenotype of endothelial cells *in vitro* and decreases vascular growth *in vivo* [73]. This suggests MALAT1 also promotes tumor growth by enhancing angiogenesis. Moreover, the fundamental roles of lncRNA in cancer cell metastasis have been intensively studied. HOTAIR is strongly expressed in breast cancer. The expression of HOTAIR can significantly predict the prognosis as well as metastatic status of breast cancer patients. Increased HOTAIR will result in breast cancer metastasis *in vivo* [74]. HOTAIR is also elevated in several cancers and contributes to their metastasis [75]. LncRNA, Lymph Node Metastasis Associated Transcript 1 (LNMAT1), is found to be up-regulated in lymph node metastasis from bladder cancer and regulates CCL2 expression epigenetically. The LNMAT1-induced CCL2 recruits tumor associated macrophages which promotes metastasis of bladder cancer through VEGF secretion [76]. Increasing numbers of metastasis-related lncRNAs are being identified. LncRNA-activated by TGF-β (lncRNA-ATB) is up-regulated in hepatocellular carcinoma metastases and promotes cancer distant metastasis by sponging miR-200 family [77]. In addition, lncRNAs also contribute to therapy resistance of cancer cells. For example, long non-coding RNA MIR100HG confers acquired cetuximab resistance by splicing into miR-100 and miR-125b

[78]. High expression of lncRNA colorectal cancer-associated lncRNA (CCAL) will predict worse response to adjuvant chemotherapy of colorectal cancer. It is found that CCAL can induce multidrug resistance through activating Wnt signaling pathway and enhancing P-gp expression [79].

TARGETING LONG NON-CODING RNA FOR CANCER THERAPY

Given the fundamental roles of lncRNAs in carcinogenesis and cancer progression, therapeutic approaches to reverse abnormal expression of lncRNAs might be promising for a clinical perspective. It is acknowledged that lncRNAs appear to be cancer specific. Targeting lncRNAs might result in fewer side effects. There are several approaches to target lncRNAs for cancer therapy. RNA interference (RNAi) which is agonaute- and dicer-dependent RNA cleavage silencing has been used to target oncogenic lncRNAs. RNase H based antisense oligonucleotides (ASOs) were also developed to knockdown lncRNAs *in vivo*. Alternatively, small molecular inhibitors which were used to target protein-protein interactions were recently found to be druggable agents against lncRNAs. Similarly, recent advances in genomic editing such as CRISPR/Cas9 system make it possible to achieve robust and consistent knockdown of lncRNAs.

RNA Interference

The RNAi pathway was first discovered in *Caenorhabditis elegans* by delivery of ectogenic double strand RNA to knockdown corresponding RNA transcripts. Since then insights into the molecular mechanism of RNAi have been uncovered. The double strand RNA is processed by Dicer into siRNA. SiRNA components then serve as sequence determinants of RISC and cleave base-complementarity mRNAs through the endonuclease argonaute protein [80]. Generation of loss-of-function phenotypes via RNAi has been implicated in functional studies of several models from Drosophila to mouse. RNAi also functions in human cells and has been studied in several clinical studies [81].

Similarly, siRNAs can abrogate cancer tumorigenesis and progression by targeting oncogenic lncRNAs. For instance, the oncogenic lncRNA focally amplified lncRNA on chromosome 1 (FAL1) contributes to ovarian cancer progression. SiRNAs targeting FAL1 significantly inhibit ovarian cancer cell proliferation [82]. SiRNAs specifically targeting PCNA-AS1, a carcinogenic antisense lncRNA which promotes hepatocellular carcinoma growth through enhancing PCNA mRNA stability inhibit cancer growth *in vitro* [83]. Metastasis of prostate cancer can also be impaired by siRNAs targeting the oncogenic lncRNA second chromosome locus associated with prostate-1 (SChLAP1) [84]. Several lncRNAs are indeed knocked down by siRNAs in different types of cancers *in vitro* [85 - 89]. However, there are still problems to be addressed for

RNAi. Unlike messenger RNAs which are often translocated into cytoplasm to be translated, the location of lncRNAs is dynamic. Several lncRNAs prefer to reside in the nucleus, some show a distribution between cytoplasm and nucleus and others tend to be accumulated in the cytoplasm [90]. As the machinery of RNAi is localized in the cytoplasm, to some extent, RNAi is more efficient to degrade lncRNAs dwelling in cytoplasm.

Antisense Oligonucleotides

ASOs are single strand nucleic acid sequences between 8-20 nt which target RNAs upon the hybridization to the sense complementary RNA. ASOs can degrade RNAs through recruitment of RNase H to the DNA-RNA heteroduplex, or inhibit the translation of mRNAs through abrogating mRNA-ribosome subunits interactions [91]. ASOs also show high efficiency in knocking down lncRNAs. The first generation of ASOs is acquired by modifying the phosphate backbone of DNA into phosphorothioate. The second generation of ASOs are formed by additional modifications of the nucleic acid at the ribose sugar. The combination of modifying phosphate, ribose sugar and nucleoside leads to achieving the best result of third generation ASOs [92]. To protect ASOs from exonucleases *in vivo*, locked nucleic acids (LNAs) modified ASOs composing of nucleic acid analogues were designed. The LNAs show lesser self complementary and increased binding affinity than other chemically modified ASOs [93]. The chemical structures of ASOs are presented in Fig. (**3**)

The application of ASOs extends from knockdown of mRNAs to target lncRNAs. LNAs targeting lncARSR (lncRNA Activated in RCC with Sunitinib Resistance) resensitize renal cell carcinoma to sunitinib and they achieve partial response in patient derived xenograft models [94]. LncRNA BCAR4 can promote breast cancer metastasis through activating non-canonical hedgehog pathway epigenetically, but LNAs targeting BCAR4 suppress breast cancer metastasis *in vivo* [95]. LNAs specifically knocking down lncRNA MALAT1 can inhibit lung cancer metastasis in xenograft models [96]. Additionally, to achieve higher hybridization ability and reduced side effects, LNAs were designed to form a LNA-DNA-LNA complex which contains a nucleotide phosphorothioate gap; hence the complex is termed as ASO gapmer. Gapmers also show good anti-lncRNA performance. Silencing of SAMMSON by gapmers disrupts mitochondria functions of melanoma cells and sensitizes melanoma to mitogen-activated protein kinase inhibitors *in vivo* [68]. Gapmers targeting MALAT1 significantly antagonize multiple myeloma cell proliferation and result in apoptosis *in vivo* and in *vitro* [69]. Gapmers can also function in thyroid cancer, bladder cancer, leukemia and several other tumors [97 - 101]. Because RNase H1 enzyme is responsible for ASO-mediated RNA degradation and RNase H1 mainly residues in the nucleus, ASOs are more efficient in knocking down nuclear lncRNAs than RNAi [102].

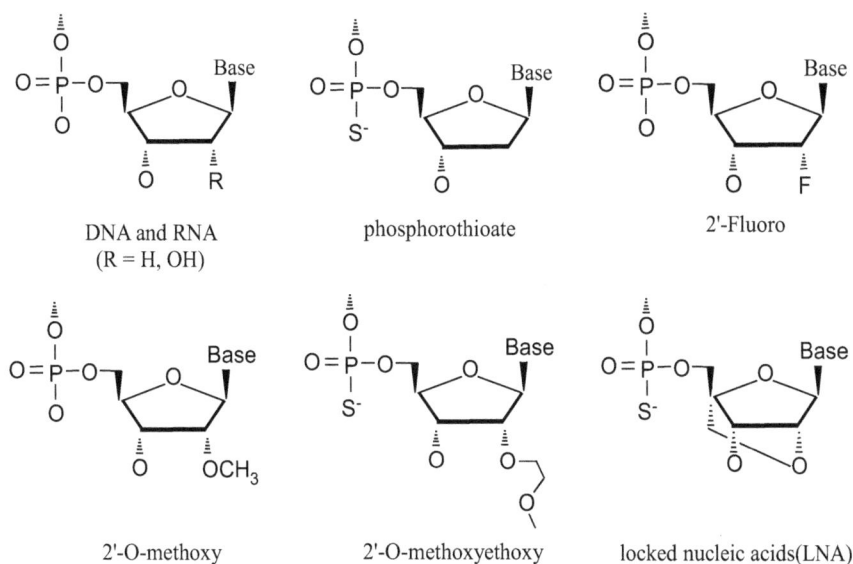

Fig. (3). Different chemical structures of ASOs. The first generations of ASOs termed as phosphorothioate are acquired by replacing the oxygen atom of the phosphate backbone of DNA with a sulfur atom. The second and third generations of ASOs are formed by modifying phosphate backbone or ribose sugar respectively. As for LNA, the ribose sugar is modified by forming chemical bond between 2' and 4' oxygen atom.

Aptamers

Aptamers are structured proteins or oligonucleotides which can recognize specific molecules by forming tertiary structures with them [103]. Many lncRNAs tend to form secondary structures which always result in the failure of RNAi or ASOs targeting. The use of aptamers may overcome this limitation, as aptamers have higher specificity and binding affinity due to their senior structures [103]. The first discovery of lncRNA-associated aptamers was lncRNA themselves. LncRNA growth arrest specific 5 (Gas5) forms hairpin structures and docks with glucocorticoid receptor via its glucocorticoid response element [104]. An lncRNA, polyadenylated nucleus RNA (PAN RNA), derived from Kaposi's sarcoma-associated herpesvirus (KSHV) can specifically interact with viral latency-associated nucleus antigen (LANA) and promote disassociations of LANA episomes during KSHV reactivation [105]. An artificial lncRNA modulated by binding sites of several different miRNAs and aptamers antagonizes β-catenin and NF-κB. It shows strong regulation on bladder cancer proliferation and apoptosis [106]. Since aptamers show very high specificity to their targets, they can be used to detect lncRNAs and serve as diagnostic tools. CG3-aptamer specifically binds to an lncRNA, prostate cancer antigen 3 (PCA3), and can stain prostate cancer as well as benign prostatic hyperplasia [107].

Chemistry and Delivery of RNA-based Therapeutics

RNA-based therapies such as siRNAs, ASOs and aptamers, have great potential in clinical practice as they abrogate the function of carcinogenetic lncRNAs. However, there are still many obstacles to overcome when applying RNA-based therapies for anti-cancer treatment. These oligonucleotides are innately unstable *in vivo* as the result of the omnipresence of different kinds of ribonucleases [108]. These RNA-based therapies can trigger the immune response [109]. So as to deliver these RNA-based therapies to their destination *in vivo* and efficiently knockdown lncRNAs, these RNA-based therapies must recognize target cells specifically, be transported across cell membranes and escape from lysosome [110 - 112].

Chemical modification is essential to enhance the stability of these RNA-based therapies without suppressing their biological functions. The oxygen atom of the phosphate backbone of DNA can be substituted for sulfur which turns oligonucleotides resistance to the nucleases, improves hydrophobicity of these oligonucleotides and increases pharmackinetices as a result of its high binding affinity to plasma proteins [113]. Other kinds of modifications of the backbone have been tried and turn out to be successful. By transforming the 2'-O-hyhroxy of the amyl sugar phosphate backbone into 2'-O-methoxy, 2'-O-methoxyethoxy or 2'-fluoro and so on, the pharmackinetices of these engineered oligonucleotides are improved [91, 108]. Interestingly, among other common modifications, ASOs modified by LNAs show the highest RNA-binding affinity to DNA. The phosphate backbone of LNAs is bicyclic ribose with a connection between 2'-O and 4'-O by a methylene bridge. The LNA-based strategies show great efficiency in knocking down the abnormal expression of lncRNA both *in vitro* and *in vivo* as discussed above.

In addition, delivery of RNA-based therapeutics to their destination can be improved by the specific conjuncted compartments. Antibodies, aptamer or synthetic peptides can be either attached to the RNA molecule directly or to the delivery carriers indirectly to enhance the specificity [114]. Antibodies and aptamers show effective binding affinity and strong interaction with their target. Aptamers show better tissue penetration but lower financial burden [108, 115]. Synthetic peptides have lower binding specificities to the membrane receptors when compared to antibodies and aptamers.

To increase the stability as well as the safety of RNA-based therapeutics, delivery systems of artificial carriers have also been applied. Viral system such as: retrovirus system, adenovirus systems and adeno-associated virus systems have been genetically recombined for delivery of RNA-based therapeutics to target lncRNAs *in vitro* and in mouse models [116 - 121]. Several clinical trials were designed to evaluate the efficacy and safety of these viral vectors [122 - 124]. Viral delivery systems indeed show success in targeting genes and might have very little vector-related clinical toxicity. However, they enjoy very high immunogenicity and will result in the emergence of anti-viral capsid specific T cells which might limit their application in clinical use. To overcome the drawback of viral vectors, non-viral delivery systems are becoming gradually accepted. Lipid-based vectors, polymeric delivery systems and inorganic materials can either encapsulate or interact with RNA to form stable nanoparticles through the electrostatic forces. These shelters protect RNA from nuclease degradation completely and also enhance the penetration as well as diffusion ability of RNA base-therapeutics by permeability and retention effects [108, 125, 126]. They show good performance in delivery of anti-lncRNA *in vivo*. For example, a liposomal carrier, phosphocholine derived nanoparticles, can transport siRNA specifically targeting lncRNA Ceruloplasmin in ovarian cancer and significantly decrease ovarian cancer growth *in vivo* [127]. However, these non-viral delivery approaches still have their drawbacks. Liposomes have limited freeze-thaw cycles and can accumulate in several organs such as the liver, spleen and lungs. Polymeric vectors are difficult to manipulate the nanoparticles size and they are cost-effective [126, 128].

Small Molecules

Small molecules have been successfully adapted to abrogate protein to protein interactions in human proteome [129 - 131]. Small molecular inhibitors can directly interact with docking pockets of proteins which, in doing so, abrogate the interaction between different proteins. They can bind to sites irrelevant to the interaction interface which always results in the alteration of protein conformation and inhibits the protein-protein interactions or dimerization. Similarly, functional RNAs exert their regulatory roles via their senior structures. Therefore, they can be targeted by small molecular drugs. For instance, small molecular compounds interact with the specific structural elements of viral RNA translation initiation region and inhibit Hepatitis C Virus as well as Human Immunodeficiency Virus replication [132]. A structural regulatory element termed as: riboswitches can form in the 5' untranslation region of mRNA in bacteria. Some small molecules are produced to target riboswitches of bacterial mRNA and decrease bacterial mRNA translation [133]. These evidences from other regulatory RNAs suggest lncRNAs might also be modulated by small molecule drugs, as lncRNAs can form secondary and tertiary structures [134 - 136]. Through coupling with the conformational structure of binding domains and thereby interacting with kinases, transcriptional factors, histones, DNA as well as RNA; lncRNAs can exert their functionally regulatory roles in cancer [137]. Targeting these interactions by small molecular drugs might be possible, as small molecular drugs can mask the binding interfaces between lncRNAs and different molecules. Camptothecin, for example, can inhibit the interaction between HOTAIR and Enhancer of zeste homolog 2 (EZH2) and thus decrease the level of HOXD11 whose expression is dependent on the HOTAIR-EZH2 interaction [138]. In addition, high-throughput screening systems have been used to screen small molecular drugs which might inhibit RNA function [139, 140]. Small molecules targeting lncRNA are still promising, although they remain to be explored and tested.

CRISPR System

Despite that many lncRNAs can be targeted by RNA-based therapeutics, it is hard to achieve constantly robust knockdown by RNA-based therapeutics. Recent advances in genetic editing tools such as: clustered regularly interspaced short palindromic repeats (CRISPR) systems transcription activation-like element nucleases (TALENs) system and zinc finger nuclease (ZFN) make it possible to target lncRNAs from transcriptional repression to knockout. Using two pairs of TALENs, it has been possible to delete the genomic region for coding MALAT1 in zebrafish [141]. ZFN system enables us to introduce destabilizing elements into lncRNA genomic regions and generates loss-of-lncRNA cancer models [142]. In addition, CRISPR is an emerging genetic editing system because it is easier to generate and target genes in genome wide sequence with more efficiency. Through using guide RNA libraries, an artificial transcriptional repressing fusion protein can be recruited at the transcriptional start sites of specific genes to induce transcriptional repression of specific genes in human [143]. In this CRISPR interference (CRISPRi) approach, a large scale screening of functional lncRNAs was carried out and 499 lncRNAs loci turned out to be necessary for cancer cell growth [144]. These results indicate that targeting lncRNA via CRISPRi is promising and can be potentially used in silencing oncogenetic lncRNAs genome wide in cancer. CRISPR system is used to delete the lncRNA promoters and achieve constant lncRNA knockout [145]. In addition to deleting genomic fragments, CRISPR system can result in the entire lncRNA deletion. A heritable deletion of lncRNA Rian whose genomic region is up to 23kb is generated by CRISPR/Cas9 system in mice [146].

Therapeutic Manipulation of LncRNAs

Aside from strategies mentioned above, there are other therapeutic approaches which might be used to manipulate lncRNAs to target cancer. An artificial construct BC-819 (DTA-H19) synthesized by H19 promoter and diphtheria toxin gene is promising [147]. The level of diphtheria toxin is specifically dependent on the action of H19 promoter and thus the high abundance of active H19 will result in the high level of diphtheria toxin. Since many cancers present high expression of H19 specifically, this construct might have little toxicity to normal cells. A phase I as well as IIa study shows that patients administrated with BC-819 receive clinical benefits including partial response, complete response and prolonged disease free survival time [148].

CONCLUSION

LncRNAs are notably involved in multiple biological processes of the tumor including cancer initiation, stemness, proliferation state and metastasis. LncRNAs exert their regulatory roles by acting in *trans*, in *cis*, post-transcriptional or post-translational modifications. This chapter offers insights into the fundamental roles of lncRNA in tumorigensis. Therapeutic approaches to reverse abnormal expression of lncRNA have been evaluated *in vitro*, *in vivo* and even in clinical trials. Targeting lncRNAs might be promising for anti-cancer therapy.

ACKNOWLEDGMENT

This work is supported by the National Key R&D Program of China (2017YFC1309103 and 2017YFC1309104); the Natural Science Foundation of China (81872139, 81672594 and 81772836); Fundamental Research Funds for the Central Universities (17ykjc13); Sun Yat-sen memorial hospital cultivation project for clinical research (SYS-C-201805). The authors also acknowledge Cancer Research Wales and Cardiff University China Medical Scholarship for supporting the study.

CONFLICT OF INTEREST

The authors declare no conflict of interest, financial or otherwise.

REFERENCE

[1]   Pertea M, Shumate A, Pertea G, *et al.* Thousands of large-scale RNA sequencing experiments yield a comprehensive new human gene list and reveal extensive transcriptional noise. bioRxiv 2018; 332825.

[2]   Ulitsky I. Evolution to the rescue: using comparative genomics to understand long non-coding RNAs. Nat Rev Genet 2016; 17(10): 601-14.
        [http://dx.doi.org/10.1038/nrg.2016.85] [PMID: 27573374]

[3]   Quinn JJ, Chang HY. Unique features of long non-coding RNA biogenesis and function. Nat Rev Genet 2016; 17(1): 47-62.
        [http://dx.doi.org/10.1038/nrg.2015.10] [PMID: 26666209]

[4]   Palazzo AF, Lee ES. Non-coding RNA: what is functional and what is junk? Front Genet 2015; 6: 2.
        [http://dx.doi.org/10.3389/fgene.2015.00002] [PMID: 25674102]

[5]   Mattick JS, Rinn JL. Discovery and annotation of long noncoding RNAs. Nat Struct Mol Biol 2015; 22(1): 5-7.
[http://dx.doi.org/10.1038/nsmb.2942] [PMID: 25565026]

[6]   Consortium F. I R G E R G P and Team I 2002 Analysis of the mouse transcriptome based on functional annotation of 60,770 full-length cDNAs Nature 2002; 420(563)

[7]   Chen L-L. Linking long noncoding RNA localization and function. Trends Biochem Sci 2016; 41(9): 761-72.
[http://dx.doi.org/10.1016/j.tibs.2016.07.003] [PMID: 27499234]

[8]   Salamon I, Saccani Jotti G, Condorelli G. The long noncoding RNA landscape in cardiovascular disease: a brief update. Curr Opin Cardiol 2018; 33(3): 282-9.
[PMID: 29543669]

[9]   Kleaveland B, Shi CY, Stefano J, Bartel DP. A Network of Noncoding Regulatory RNAs Acts in the Mammalian Brain. Cell 2018; 174(2): 350-362.e17.
[http://dx.doi.org/10.1016/j.cell.2018.05.022] [PMID: 29887379]

[10]   Mirza AH, Kaur S, Pociot F. Long non-coding RNAs as novel players in β cell function and type 1 diabetes. Hum Genomics 2017; 11(1): 17.
[http://dx.doi.org/10.1186/s40246-017-0113-7] [PMID: 28738846]

[11]   Anderson DM, Anderson KM, Chang C-L, *et al.* A micropeptide encoded by a putative long noncoding RNA regulates muscle performance. Cell 2015; 160(4): 595-606.
[http://dx.doi.org/10.1016/j.cell.2015.01.009] [PMID: 25640239]

[12]   Khorkova O, Hsiao J, Wahlestedt C. Basic biology and therapeutic implications of lncRNA. Adv Drug Deliv Rev 2015; 87: 15-24.
[http://dx.doi.org/10.1016/j.addr.2015.05.012] [PMID: 26024979]

[13]   St Laurent G, Wahlestedt C, Kapranov P. The Landscape of long noncoding RNA classification. Trends Genet 2015; 31(5): 239-51.
[http://dx.doi.org/10.1016/j.tig.2015.03.007] [PMID: 25869999]

[14]   Wei Z, Smith JA, Mao G, *et al.* The cis and trans effects of the risk variants of coronary artery disease in the Chr9p21 region. BMC Med Genomics 2015; 8: 1-12.
[PMID: 25582225]

[15]   Derrien T, Johnson R, Bussotti G, *et al.* The GENCODE v7 catalog of human long noncoding RNAs: analysis of their gene structure, evolution, and expression. Genome Res 2012; 22(9): 1775-89.
[http://dx.doi.org/10.1101/gr.132159.111] [PMID: 22955988]

[16]   Wu X, Bartel DP. Widespread Influence of 3′-End Structures on Mammalian mRNA Processing and Stability. Cell 2017; 169(5): 905-917.e11.
[http://dx.doi.org/10.1016/j.cell.2017.04.036] [PMID: 28525757]

[17]   Zhang B, Gunawardane L, Niazi F, Jahanbani F, Chen X, Valadkhan S. A novel RNA motif mediates the strict nuclear localization of a long noncoding RNA. Mol Cell Biol 2014; 34(12): 2318-29.
[http://dx.doi.org/10.1128/MCB.01673-13] [PMID: 24732794]

[18]   Sakaguchi T, Hasegawa Y, Brockdorff N, *et al.* Control of Chromosomal Localization of Xist by hnRNP U Family Molecules. Dev Cell 2016; 39(1): 11-2.
[http://dx.doi.org/10.1016/j.devcel.2016.09.022] [PMID: 27728779]

[19]   Hacisuleyman E, Goff LA, Trapnell C, *et al.* Topological organization of multichromosomal regions by the long intergenic noncoding RNA Firre. Nat Struct Mol Biol 2014; 21(2): 198-206.
[http://dx.doi.org/10.1038/nsmb.2764] [PMID: 24463464]

[20]   Blythe AJ, Fox AH, Bond CS. The ins and outs of lncRNA structure: How, why and what comes next? Biochim Biophys Acta 2016; 1859(1): 46-58.
[http://dx.doi.org/10.1016/j.bbagrm.2015.08.009] [PMID: 26325022]

[21] Crowther C V, Jones L E, Morelli J N, *et al.* Influence of Two Bulge Loops on the Stability of RNA Duplexes Rna-a Publication of the Rna Society 2016; 23 rna.056168.116

[22] Mondal T, Subhash S, Vaid R, *et al.* MEG3 long noncoding RNA regulates the TGF-β pathway genes through formation of RNA-DNA triplex structures. Nat Commun 2015; 6: 7743.
[http://dx.doi.org/10.1038/ncomms8743] [PMID: 26205790]

[23] Novikova IV, Hennelly SP, Sanbonmatsu KY. Structural architecture of the human long non-coding RNA, steroid receptor RNA activator. Nucleic Acids Res 2012; 40(11): 5034-51.
[http://dx.doi.org/10.1093/nar/gks071] [PMID: 22362738]

[24] Wu H, Yang L, Chen LL. Diversity of Long Noncoding RNAs and Their Generation Trends in Genetics Tig 2017; 33: S0168952517300859.

[25] Schmitt AM, Chang HY. Long Noncoding RNAs: At the Intersection of Cancer and Chromatin Biology. Cold Spring Harb Perspect Med 2017; 7(7): a026492.
[http://dx.doi.org/10.1101/cshperspect.a026492] [PMID: 28193769]

[26] Malek R, Gajula RP, Williams RD, *et al.* TWIST1-WDR5-*Hottip* Regulates *Hoxa9* Chromatin to Facilitate Prostate Cancer Metastasis. Cancer Res 2017; 77(12): 3181-93.
[http://dx.doi.org/10.1158/0008-5472.CAN-16-2797] [PMID: 28484075]

[27] Trimarchi T, Bilal E, Ntziachristos P, *et al.* Genome-wide mapping and characterization of Notch-regulated long noncoding RNAs in acute leukemia. Cell 2014; 158(3): 593-606.
[http://dx.doi.org/10.1016/j.cell.2014.05.049] [PMID: 25083870]

[28] Xiang JF, Yin QF, Chen T, *et al.* Human colorectal cancer-specific CCAT1-L lncRNA regulates long-range chromatin interactions at the MYC locus. Cell Res 2014; 24(5): 513-31.
[http://dx.doi.org/10.1038/cr.2014.35] [PMID: 24662484]

[29] Lai F, Orom UA, Cesaroni M, *et al.* Activating RNAs associate with Mediator to enhance chromatin architecture and transcription. Nature 2013; 494(7438): 497-501.
[http://dx.doi.org/10.1038/nature11884] [PMID: 23417068]

[30] Latgã© G, Poulet C, Bours V, Josse C, Jerusalem G. Natural Antisense Transcripts: Molecular Mechanisms and Implications in Breast Cancers International Journal of Molecular Sciences 2018; 19: 123.

[31] Beckedorff FC, Ayupe AC, Crocci-Souza R, *et al.* The intronic long noncoding RNA ANRASSF1 recruits PRC2 to the RASSF1A promoter, reducing the expression of RASSF1A and increasing cell proliferation. PLoS Genet 2013; 9(8): e1003705.
[http://dx.doi.org/10.1371/journal.pgen.1003705] [PMID: 23990798]

[32] Delmonico L, Moreira AdosS, Franco MF, *et al.* CDKN2A (p14(ARF)/p16(INK4a)) and ATM promoter methylation in patients with impalpable breast lesions. Hum Pathol 2015; 46(10): 1540-7.
[http://dx.doi.org/10.1016/j.humpath.2015.06.016] [PMID: 26255234]

[33] Meseure D, Vacher S, Alsibai KD, *et al.* Expression of ANRIL-Polycomb Complexes-CDKN2A/B/ARF Genes in Breast Tumors: Identification of a Two-Gene (EZH2/CBX7) Signature with Independent Prognostic Value. Mol Cancer Res 2016; 14(7): 623-33.
[http://dx.doi.org/10.1158/1541-7786.MCR-15-0418] [PMID: 27102007]

[34] Marcho C, Bevilacqua A, Tremblay KD, Mager J. Tissue-specific regulation of Igf2r/Airn imprinting during gastrulation. Epigenetics Chromatin 2015; 8: 10.
[http://dx.doi.org/10.1186/s13072-015-0003-y] [PMID: 25918552]

[35] Mallo M. Reassessing the Role of Hox Genes during Vertebrate Development and Evolution Trends in Genetics Tig 2017; 34 S0168952517302093

[36] Deckard CE, Sczepanski JT. Polycomb repressive complex 2 binds RNA irrespective of stereochemistry. Chem Commun (Camb) 2018; 54(85): 12061-4.
[http://dx.doi.org/10.1039/C8CC07433J] [PMID: 30295686]

[37]   Bao X, Wu H, Zhu X, *et al.* The p53-induced lincRNA-p21 derails somatic cell reprogramming by sustaining H3K9me3 and CpG methylation at pluripotency gene promoters. Cell Res 2015; 25(1): 80-92.
[http://dx.doi.org/10.1038/cr.2014.165] [PMID: 25512341]

[38]   Williams GR. Cloning and characterization of two novel thyroid hormone receptor β isoforms. Mol Cell Biol 2000; 20(22): 8329-42.
[http://dx.doi.org/10.1128/MCB.20.22.8329-8342.2000] [PMID: 11046130]

[39]   Hastings ML, Wilson CM, Munroe SH. A purine-rich intronic element enhances alternative splicing of thyroid hormone receptor mRNA Rna-a Publication of the Rna Society. 2001; 7: pp. 859-74.

[40]   Zhang X, Hamblin MH, Yin KJ. The long noncoding RNA Malat1: Its physiological and pathophysiological functions. RNA Biol 2017; 14(12): 1705-14.
[http://dx.doi.org/10.1080/15476286.2017.1358347] [PMID: 28837398]

[41]   Fox AH, Nakagawa S, Hirose T, Bond CS. Paraspeckles: Where Long Noncoding RNA Meets Phase Separation. Trends Biochem Sci 2018; 43(2): 124-35.
[http://dx.doi.org/10.1016/j.tibs.2017.12.001] [PMID: 29289458]

[42]   Yamazaki T, Souquere S, Chujo T, *et al.* Functional Domains of NEAT1 Architectural lncRNA Induce Paraspeckle Assembly through Phase Separation. Mol Cell 2018; 70(6): 1038-1053.e7.
[http://dx.doi.org/10.1016/j.molcel.2018.05.019] [PMID: 29932899]

[43]   Jiang L, Shao C, Wu QJ, *et al.* NEAT1 scaffolds RNA-binding proteins and the Microprocessor to globally enhance pri-miRNA processing. Nat Struct Mol Biol 2017; 24(10): 816-24.
[http://dx.doi.org/10.1038/nsmb.3455] [PMID: 28846091]

[44]   Chen J, Shen B. RNA Bioinformatics for Precision Medicine. Adv Exp Med Biol 2016; 939: 21-38.
[http://dx.doi.org/10.1007/978-981-10-1503-8_2] [PMID: 27807742]

[45]   Matoulkova E, Michalova E, Vojtesek B, Hrstka R. The role of the 3′ untranslated region in post-transcriptional regulation of protein expression in mammalian cells. RNA Biol 2012; 9(5): 563-76.
[http://dx.doi.org/10.4161/rna.20231] [PMID: 22614827]

[46]   Gong C, Maquat LE. lncRNAs transactivate STAU1-mediated mRNA decay by duplexing with 3′ UTRs via Alu elements. Nature 2011; 470(7333): 284-8.
[http://dx.doi.org/10.1038/nature09701] [PMID: 21307942]

[47]   Kretz M, Siprashvili Z, Chu C, *et al.* Control of somatic tissue differentiation by the long non-coding RNA TINCR. Nature 2013; 493(7431): 231-5.
[http://dx.doi.org/10.1038/nature11661] [PMID: 23201690]

[48]   Taniguchi K, Karin M. NF-κB, inflammation, immunity and cancer: coming of age. Nat Rev Immunol 2018; 18(5): 309-24.
[http://dx.doi.org/10.1038/nri.2017.142] [PMID: 29379212]

[49]   Liu B, Sun L, Liu Q, *et al.* A cytoplasmic NF-κB interacting long noncoding RNA blocks IκB phosphorylation and suppresses breast cancer metastasis. Cancer Cell 2015; 27(3): 370-81.
[http://dx.doi.org/10.1016/j.ccell.2015.02.004] [PMID: 25759022]

[50]   Wang P, Xue Y, Han Y, *et al.* The STAT3-binding long noncoding RNA lnc-DC controls human dendritic cell differentiation. Science 2014; 344(6181): 310-3.
[http://dx.doi.org/10.1126/science.1251456] [PMID: 24744378]

[51]   Yang F, Zhang H, Mei Y, Wu M. Reciprocal regulation of HIF-1α and lincRNA-p21 modulates the Warburg effect. Mol Cell 2014; 53(1): 88-100.
[http://dx.doi.org/10.1016/j.molcel.2013.11.004] [PMID: 24316222]

[52]   Yu X, Zheng H, Chan MT, Wu WK. HULC: an oncogenic long non-coding RNA in human cancer. J Cell Mol Med 2017; 21(2): 410-7.
[http://dx.doi.org/10.1111/jcmm.12956] [PMID: 27781386]

[53]   Cesana M, Cacchiarelli D, Legnini I, *et al.* A long noncoding RNA controls muscle differentiation by functioning as a competing endogenous RNA. Cell 2011; 147(2): 358-69.
[http://dx.doi.org/10.1016/j.cell.2011.09.028] [PMID: 22000014]

[54]   Li Z, Jiang P, Li J, *et al.* Tumor-derived exosomal lnc-Sox2ot promotes EMT and stemness by acting as a ceRNA in pancreatic ductal adenocarcinoma. Oncogene 2018; 37(28): 3822-38.
[http://dx.doi.org/10.1038/s41388-018-0237-9] [PMID: 29643475]

[55]   Liang L, Xu J, Wang M, *et al.* LncRNA HCP5 promotes follicular thyroid carcinoma progression via miRNAs sponge. Cell Death Dis 2018; 9(3): 372.
[http://dx.doi.org/10.1038/s41419-018-0382-7] [PMID: 29515098]

[56]   Li M, Xie Z, Wang P, *et al.* The long noncoding RNA GAS5 negatively regulates the adipogenic differentiation of MSCs by modulating the miR-18a/CTGF axis as a ceRNA. Cell Death Dis 2018; 9(5): 554.
[http://dx.doi.org/10.1038/s41419-018-0627-5] [PMID: 29748618]

[57]   He S, Su H, Liu C, *et al.* MicroRNA-encoding long non-coding RNAs. BMC Genomics 2008; 9: 236.
[http://dx.doi.org/10.1186/1471-2164-9-236] [PMID: 18492288]

[58]   Dey BK, Pfeifer K, Dutta A. The H19 long noncoding RNA gives rise to microRNAs miR-675-3p and miR-675-5p to promote skeletal muscle differentiation and regeneration. Genes Dev 2014; 28(5): 491-501.
[http://dx.doi.org/10.1101/gad.234419.113] [PMID: 24532688]

[59]   Legnini I, Morlando M, Mangiavacchi A, Fatica A, Bozzoni I. A feedforward regulatory loop between HuR and the long noncoding RNA linc-MD1 controls early phases of myogenesis. Mol Cell 2014; 53(3): 506-14.
[http://dx.doi.org/10.1016/j.molcel.2013.12.012] [PMID: 24440503]

[60]   Flavahan W A, Gaskell E, Bernstein B E. 2017; Epigenetic plasticity and the hallmarks of cancer Science 357: eaal2380.
[http://dx.doi.org/10.1126/science.aal2380]

[61]   Yan P, Luo S, Lu JY, Shen X. Cis- and trans-acting lncRNAs in pluripotency and reprogramming. Curr Opin Genet Dev 2017; 46: 170-8.
[http://dx.doi.org/10.1016/j.gde.2017.07.009] [PMID: 28843809]

[62]   Lin N, Chang K-Y, Li Z, *et al.* An evolutionarily conserved long noncoding RNA TUNA controls pluripotency and neural lineage commitment. Mol Cell 2014; 53(6): 1005-19.
[http://dx.doi.org/10.1016/j.molcel.2014.01.021] [PMID: 24530304]

[63]   Li H, Zhu L, Xu L, *et al.* Long noncoding RNA linc00617 exhibits oncogenic activity in breast cancer. Mol Carcinog 2017; 56(1): 3-17.
[http://dx.doi.org/10.1002/mc.22338] [PMID: 26207516]

[64]   Pingping Z, Jiayi W, Yanying W, *et al.* LncGata6 maintains stemness of intestinal stem cells and promotes intestinal tumorigenesis Nature Cell Biology

[65]   Li M, Chai H F, Peng F, *et al.* Estrogen receptor β upregulated by lncRNA-H19 to promote cancer stem-like properties in papillary thyroid carcinoma Cell Death and Disease
[http://dx.doi.org/10.1038/s41419-018-1077-9]

[66]   Jing R, Liang D, Zhang D, *et al.* Carcinoma-associated fibroblasts promote the stemness and chemoresistance of colorectal cancer by transferring exosomal lncRNA H19 Theranostics 2018; 8: 3932-48.

[67]   White NM, Zhao SG, Zhang J, *et al.* Multi-institutional Analysis Shows that Low PCAT-14 Expression Associates with Poor Outcomes in Prostate Cancer. Eur Urol 2017; 71(2): 257-66.
[http://dx.doi.org/10.1016/j.eururo.2016.07.012] [PMID: 27460352]

[68]   Leucci E, Vendramin R, Spinazzi M, *et al.* Melanoma addiction to the long non-coding RNA

SAMMSON. Nature 2016; 531(7595): 518-22.
[http://dx.doi.org/10.1038/nature17161] [PMID: 27008969]

[69]   Amodio N, Stamato MA, Juli G, *et al.* Drugging the lncRNA MALAT1 via LNA gapmeR ASO inhibits gene expression of proteasome subunits and triggers anti-multiple myeloma activity. Leukemia 2018; 32(9): 1948-57.
[http://dx.doi.org/10.1038/s41375-018-0067-3] [PMID: 29487387]

[70]   Malakar P, Shilo A, Mogilevsky A, *et al.* Long Noncoding RNA MALAT1 Promotes Hepatocellular Carcinoma Development by SRSF1 Upregulation and mTOR Activation. Cancer Res 2017; 77(5): 1155-67.
[http://dx.doi.org/10.1158/0008-5472.CAN-16-1508] [PMID: 27993818]

[71]   Huang JK, Ma L, Song WH, *et al.* MALAT1 promotes the proliferation and invasion of thyroid cancer cells via regulating the expression of IQGAP1. Biomed Pharmacother 2016; 83: 1-7.
[http://dx.doi.org/10.1016/j.biopha.2016.05.039] [PMID: 27470543]

[72]   Yang MH, Hu ZY, Xu C, *et al.* MALAT1 promotes colorectal cancer cell proliferation/migration/invasion via PRKA kinase anchor protein 9. Biochim Biophys Acta 2015; 1852(1): 166-74.
[http://dx.doi.org/10.1016/j.bbadis.2014.11.013] [PMID: 25446987]

[73]   Huang JK, Ma L, Song WH, *et al.* LncRNA-MALAT1 promotes angiogenesis of thyroid cancer by modulating tumor-associated macrophage FGF2 protein secretion. J Cell Biochem 2017; 118(12): 4821-30.
[http://dx.doi.org/10.1002/jcb.26153] [PMID: 28543663]

[74]   Hajjari M, Salavaty A. HOTAIR: an oncogenic long non-coding RNA in different cancers. Cancer Biol Med 2015; 12(1): 1-9.
[PMID: 25859406]

[75]   Sun Z, XY W, CL W. The association between LncRNA HOTAIR and cancer lymph node metastasis and distant metastasis: a meta-analysis Neoplasma 2018; 65: 178-84.

[76]   Chen C, He W, Huang J, *et al.* LNMAT1 promotes lymphatic metastasis of bladder cancer *via* CCL2 dependent macrophage recruitment Nature Communications

[77]   Yuan JH, Yang F, Wang F, *et al.* A long noncoding RNA activated by TGF-β promotes the invasion-metastasis cascade in hepatocellular carcinoma. Cancer Cell 2014; 25(5): 666-81.
[http://dx.doi.org/10.1016/j.ccr.2014.03.010] [PMID: 24768205]

[78]   Lu Y, Zhao X, Liu Q, *et al.* lncRNA MIR100HG-derived miR-100 and miR-125b mediate cetuximab resistance via Wnt/β-catenin signaling Nature Medicine 2017; 23: 1331.

[79]   Ma Y, Yang Y, Wang F, *et al.* Long non-coding RNA CCAL regulates colorectal cancer progression by activating Wnt/β-catenin signalling pathway via suppression of activator protein 2α. Gut 2016; 65(9): 1494-504.
[http://dx.doi.org/10.1136/gutjnl-2014-308392] [PMID: 25994219]

[80]   Brummelkamp TR, Bernards R, Agami R. A system for stable expression of short interfering RNAs in mammalian cells. Science 2002; 296(5567): 550-3.
[http://dx.doi.org/10.1126/science.1068999] [PMID: 11910072]

[81]   Adams D, Suhr OB, Dyck PJ, *et al.* Trial design and rationale for APOLLO, a Phase 3, placebo-controlled study of patisiran in patients with hereditary ATTR amyloidosis with polyneuropathy. BMC Neurol 2017; 17(1): 181.
[http://dx.doi.org/10.1186/s12883-017-0948-5] [PMID: 28893208]

[82]   Hu X, Feng Y, Zhang D, *et al.* A functional genomic approach identifies FAL1 as an oncogenic long noncoding RNA that associates with BMI1 and represses p21 expression in cancer. Cancer Cell 2014; 26(3): 344-57.
[http://dx.doi.org/10.1016/j.ccr.2014.07.009] [PMID: 25203321]

[83] Yuan SX, Tao QF, Wang J, *et al.* Antisense long non-coding RNA PCNA-AS1 promotes tumor growth by regulating proliferating cell nuclear antigen in hepatocellular carcinoma. Cancer Lett 2014; 349(1): 87-94.
[http://dx.doi.org/10.1016/j.canlet.2014.03.029] [PMID: 24704293]

[84] Prensner JR, Iyer MK, Sahu A, *et al.* The long noncoding RNA SChLAP1 promotes aggressive prostate cancer and antagonizes the SWI/SNF complex. Nat Genet 2013; 45(11): 1392-8.
[http://dx.doi.org/10.1038/ng.2771] [PMID: 24076601]

[85] Deng X, Feng N, Zheng M, *et al.* PM$_{2.5}$ exposure-induced autophagy is mediated by lncRNA loc146880 which also promotes the migration and invasion of lung cancer cells. Biochim Biophys Acta, Gen Subj 2017; 1861(2): 112-25.
[http://dx.doi.org/10.1016/j.bbagen.2016.11.009] [PMID: 27836757]

[86] Han P, Li JW, Zhang BM, *et al.* The lncRNA CRNDE promotes colorectal cancer cell proliferation and chemoresistance via miR-181a-5p-mediated regulation of Wnt/β-catenin signaling. Mol Cancer 2017; 16(1): 9.
[http://dx.doi.org/10.1186/s12943-017-0583-1] [PMID: 28086904]

[87] Huang JZ, Chen M, Chen , *et al.* A Peptide Encoded by a Putative lncRNA HOXB-AS3 Suppresses Colon Cancer Growth Molecular Cell 2017; 68: 171-84.

[88] Koirala P, Huang J, Ho TT, Wu F, Ding X, Mo YY. LncRNA AK023948 is a positive regulator of AKT. Nat Commun 2017; 8: 14422.
[http://dx.doi.org/10.1038/ncomms14422] [PMID: 28176758]

[89] Li C, Wang S, Xing Z, *et al.* A ROR1-HER3-lncRNA signalling axis modulates the Hippo-YAP pathway to regulate bone metastasis. Nat Cell Biol 2017; 19(2): 106-19.
[http://dx.doi.org/10.1038/ncb3464] [PMID: 28114269]

[90] Cabili MN, Dunagin MC, McClanahan PD, *et al.* Localization and abundance analysis of human lncRNAs at single-cell and single-molecule resolution. Genome Biol 2015; 16: 20.
[http://dx.doi.org/10.1186/s13059-015-0586-4] [PMID: 25630241]

[91] Bennett CF, Swayze EE. RNA targeting therapeutics: molecular mechanisms of antisense oligonucleotides as a therapeutic platform. Annu Rev Pharmacol Toxicol 2010; 50: 259-93.
[http://dx.doi.org/10.1146/annurev.pharmtox.010909.105654] [PMID: 20055705]

[92] Goyal N, Narayanaswami P. Making sense of antisense oligonucleotides: A narrative review. Muscle Nerve 2018; 57(3): 356-70.
[http://dx.doi.org/10.1002/mus.26001] [PMID: 29105153]

[93] Watts JK. Locked nucleic acid: tighter is different. Chem Commun (Camb) 2013; 49(50): 5618-20.
[http://dx.doi.org/10.1039/c3cc40340h] [PMID: 23682352]

[94] Qu L, Ding J, Chen C, *et al.* Exosome-Transmitted lncARSR Promotes Sunitinib Resistance in Renal Cancer by Acting as a Competing Endogenous RNA. Cancer Cell 2016; 29(5): 653-68.
[http://dx.doi.org/10.1016/j.ccell.2016.03.004] [PMID: 27117758]

[95] Xing Z, Lin A, Li C, *et al.* lncRNA directs cooperative epigenetic regulation downstream of chemokine signals. Cell 2014; 159(5): 1110-25.
[http://dx.doi.org/10.1016/j.cell.2014.10.013] [PMID: 25416949]

[96] Gutschner T, Hämmerle M, Eissmann M, *et al.* The noncoding RNA MALAT1 is a critical regulator of the metastasis phenotype of lung cancer cells. Cancer Res 2013; 73(3): 1180-9.
[http://dx.doi.org/10.1158/0008-5472.CAN-12-2850] [PMID: 23243023]

[97] Seitz AK, Christensen LL, Christensen E, *et al.* Profiling of long non-coding RNAs identifies LINC00958 and LINC01296 as candidate oncogenes in bladder cancer. Sci Rep 2017; 7(1): 395.
[http://dx.doi.org/10.1038/s41598-017-00327-0] [PMID: 28341852]

[98] Schultheiss C S, Laggai S, Czepukojc B, *et al.* The long non-coding RNA H19 suppresses

carcinogenesis and chemoresistance in hepatocellular carcinoma 2017.
[http://dx.doi.org/10.15698/cst2017.10.105]

[99] Diermeier SD, Chang KC, Freier SM, *et al.* Mammary Tumor-Associated RNAs Impact Tumor Cell Proliferation, Invasion, and Migration. Cell Reports 2016; 17(1): 261-74.
[http://dx.doi.org/10.1016/j.celrep.2016.08.081] [PMID: 27681436]

[100] Kim D, Lee WK, Jeong S, *et al.* Upregulation of long noncoding RNA LOC100507661 promotes tumor aggressiveness in thyroid cancer. Mol Cell Endocrinol 2016; 431: 36-45.
[http://dx.doi.org/10.1016/j.mce.2016.05.002] [PMID: 27151833]

[101] Booy EP, McRae EK, Koul A, Lin F, McKenna SA. The long non-coding RNA BC200 (BCYRN1) is critical for cancer cell survival and proliferation. Mol Cancer 2017; 16(1): 109.
[http://dx.doi.org/10.1186/s12943-017-0679-7] [PMID: 28651607]

[102] Lennox KA, Behlke MA. Cellular localization of long non-coding RNAs affects silencing by RNAi more than by antisense oligonucleotides. Nucleic Acids Res 2016; 44(2): 863-77.
[http://dx.doi.org/10.1093/nar/gkv1206] [PMID: 26578588]

[103] Vitiello M, Tuccoli A, Poliseno L. Long non-coding RNAs in cancer: implications for personalized therapy. Cell Oncol (Dordr) 2015; 38(1): 17-28.
[http://dx.doi.org/10.1007/s13402-014-0180-x] [PMID: 25113790]

[104] Kino T, Hurt DE, Ichijo T, Nader N, Chrousos GP. Noncoding RNA gas5 is a growth arrest- and starvation-associated repressor of the glucocorticoid receptor. Sci Signal 2010; 3(107): ra8.
[http://dx.doi.org/10.1126/scisignal.2000568] [PMID: 20124551]

[105] Campbell M, Kim KY, Chang PC, *et al.* A lytic viral long noncoding RNA modulates the function of a latent protein. J Virol 2014; 88(3): 1843-8.
[http://dx.doi.org/10.1128/JVI.03251-13] [PMID: 24257619]

[106] Xie H, Zhan H, Gao Q, *et al.* Synthetic artificial "long non-coding RNAs" targeting oncogenic microRNAs and transcriptional factors inhibit malignant phenotypes of bladder cancer cells. Cancer Lett 2018; 422: 94-106.
[http://dx.doi.org/10.1016/j.canlet.2018.02.038] [PMID: 29501702]

[107] Marangoni K, Neves AF, Rocha RM, *et al.* Prostate-specific RNA aptamer: promising nucleic acid antibody-like cancer detection. Sci Rep 2015; 5: 12090.
[http://dx.doi.org/10.1038/srep12090] [PMID: 26174796]

[108] Slaby O, Laga R, Sedlacek O. Therapeutic targeting of non-coding RNAs in cancer. Biochem J 2017; 474(24): 4219-51.
[http://dx.doi.org/10.1042/BCJ20170079] [PMID: 29242381]

[109] Meng Z, Lu M. RNA Interference-Induced Innate Immunity, Off-Target Effect, or Immune Adjuvant? Front Immunol 2017; 8: 331.
[http://dx.doi.org/10.3389/fimmu.2017.00331] [PMID: 28386261]

[110] McErlean EM, McCrudden CM, McCarthy HO. Delivery of nucleic acids for cancer gene therapy: overcoming extra- and intra-cellular barriers. Ther Deliv 2016; 7(9): 619-37.
[http://dx.doi.org/10.4155/tde-2016-0049] [PMID: 27582234]

[111] Juliano RL. The delivery of therapeutic oligonucleotides. Nucleic Acids Res 2016; 44(14): 6518-48.
[http://dx.doi.org/10.1093/nar/gkw236] [PMID: 27084936]

[112] Dowdy SF. Overcoming cellular barriers for RNA therapeutics. Nat Biotechnol 2017; 35(3): 222-9.
[http://dx.doi.org/10.1038/nbt.3802] [PMID: 28244992]

[113] Eckstein F. Phosphorothioates, essential components of therapeutic oligonucleotides. Nucleic Acid Ther 2014; 24(6): 374-87.
[http://dx.doi.org/10.1089/nat.2014.0506] [PMID: 25353652]

[114] Li Z, Rana TM. Therapeutic targeting of microRNAs: current status and future challenges. Nat Rev

Drug Discov 2014; 13(8): 622-38.
[http://dx.doi.org/10.1038/nrd4359] [PMID: 25011539]

[115]   Matsui M, Corey DR. Non-coding RNAs as drug targets. Nat Rev Drug Discov 2017; 16(3): 167-79.
[http://dx.doi.org/10.1038/nrd.2016.117] [PMID: 27444227]

[116]   Li P, Ruan X, Yang L, *et al.* A liver-enriched long non-coding RNA, lncLSTR, regulates systemic
lipid metabolism in mice. Cell Metab 2015; 21(3): 455-67.
[http://dx.doi.org/10.1016/j.cmet.2015.02.004] [PMID: 25738460]

[117]   Yang F, Huo XS, Yuan SX, *et al.* Repression of the long noncoding RNA-LET by histone deacetylase
3 contributes to hypoxia-mediated metastasis. Mol Cell 2013; 49(6): 1083-96.
[http://dx.doi.org/10.1016/j.molcel.2013.01.010] [PMID: 23395002]

[118]   Zhu M, Chen Q, Liu X, *et al.* lncRNA H19/miR-675 axis represses prostate cancer metastasis by
targeting TGFBI. FEBS J 2014; 281(16): 3766-75.
[http://dx.doi.org/10.1111/febs.12902] [PMID: 24988946]

[119]   Ruan X, Li P, Cangelosi A, Yang L, Cao H. A Long Non-coding RNA, lncLGR, Regulates Hepatic
Glucokinase Expression and Glycogen Storage during Fasting. Cell Reports 2016; 14(8): 1867-75.
[http://dx.doi.org/10.1016/j.celrep.2016.01.062] [PMID: 26904944]

[120]   Chen F, Li Y, Feng Y, He X, Wang L. Evaluation of Antimetastatic Effect of lncRNA-ATB siRNA
Delivered Using Ultrasound-Targeted Microbubble Destruction. DNA Cell Biol 2016; 35(8): 393-7.
[http://dx.doi.org/10.1089/dna.2016.3254] [PMID: 27027475]

[121]   Meola N, Pizzo M, Alfano G, Surace EM, Banfi S. The long noncoding RNA Vax2os1 controls the
cell cycle progression of photoreceptor progenitors in the mouse retina Rna-a Publication of the Rna
Society. 2012; 18: pp. 111-23.

[122]   Kim K H, Dmitriev I, O'Malley J P, *et al.* A phase I clinical trial of Ad5.SSTR/TK.RGD, a novel
infectivity-enhanced bicistronic adenovirus, in patients with recurrent gynecologic cancer Clinical
Cancer Research An Official Journal of the American Association for Cancer Research 2012; 18:
3440.

[123]   Nathwani AC, Tuddenham EGD, Rangarajan S, *et al.* Adenovirus-associated virus vector-mediated
gene transfer in hemophilia B. N Engl J Med 2011; 365(25): 2357-65.
[http://dx.doi.org/10.1056/NEJMoa1108046] [PMID: 22149959]

[124]   Derek O, Amundson K K, Fernando L E, *et al.* Brain tumor eradication and prolonged survival from
intratumoral conversion of 5-fluorocytosine to 5-fluorouracil using a nonlytic retroviral replicating
vector 2012; 145-59.

[125]   Maeda H, Nakamura H, Fang J. The EPR effect for macromolecular drug delivery to solid tumors:
Improvement of tumor uptake, lowering of systemic toxicity, and distinct tumor imaging in vivo. Adv
Drug Deliv Rev 2013; 65(1): 71-9.
[http://dx.doi.org/10.1016/j.addr.2012.10.002] [PMID: 23088862]

[126]   Lavorgna G, Vago R, Sarmini M, Montorsi F, Salonia A, Bellone M. Long non-coding RNAs as novel
therapeutic targets in cancer. Pharmacol Res 2016; 110: 131-8.
[http://dx.doi.org/10.1016/j.phrs.2016.05.018] [PMID: 27210721]

[127]   Rupaimoole R, Lee J, Haemmerle M, *et al.* Long Noncoding RNA Ceruloplasmin Promotes Cancer
Growth by Altering Glycolysis. Cell Reports 2015; 13(11): 2395-402.
[http://dx.doi.org/10.1016/j.celrep.2015.11.047] [PMID: 26686630]

[128]   Shu Y, Pi F, Sharma A, *et al.* Stable RNA nanoparticles as potential new generation drugs for cancer
therapy. Adv Drug Deliv Rev 2014; 66: 74-89.
[http://dx.doi.org/10.1016/j.addr.2013.11.006] [PMID: 24270010]

[129]   Arkin MR, Tang Y, Wells JA. Small-molecule inhibitors of protein-protein interactions: progressing
toward the reality. Chem Biol 2014; 21(9): 1102-14.
[http://dx.doi.org/10.1016/j.chembiol.2014.09.001] [PMID: 25237857]

[130] Nero TL, Morton CJ, Holien JK, Wielens J, Parker MW. Oncogenic protein interfaces: small molecules, big challenges. Nat Rev Cancer 2014; 14(4): 248-62.
[http://dx.doi.org/10.1038/nrc3690] [PMID: 24622521]

[131] Roskoski R Jr. Src protein-tyrosine kinase structure, mechanism, and small molecule inhibitors. Pharmacol Res 2015; 94: 9-25.
[http://dx.doi.org/10.1016/j.phrs.2015.01.003] [PMID: 25662515]

[132] Carnevali M, Parsons J, Wyles DL, Hermann T. A modular approach to synthetic RNA binders of the hepatitis C virus internal ribosome entry site. ChemBioChem 2010; 11(10): 1364-7.
[http://dx.doi.org/10.1002/cbic.201000177] [PMID: 20564282]

[133] Orac CM, Zhou S, Means JA, Boehm D, Bergmeier SC, Hines JV. Synthesis and stereospecificity of 4,5-disubstituted oxazolidinone ligands binding to T-box riboswitch RNA. J Med Chem 2011; 54(19): 6786-95.
[http://dx.doi.org/10.1021/jm2006904] [PMID: 21812425]

[134] Novikova IV, Hennelly SP, Sanbonmatsu KY. Sizing up long non-coding RNAs: do lncRNAs have secondary and tertiary structure? Bioarchitecture 2012; 2(6): 189-99.
[http://dx.doi.org/10.4161/bioa.22592] [PMID: 23267412]

[135] Somarowthu S, Legiewicz M, Chillón I, Marcia M, Liu F, Pyle AM. HOTAIR forms an intricate and modular secondary structure. Mol Cell 2015; 58(2): 353-61.
[http://dx.doi.org/10.1016/j.molcel.2015.03.006] [PMID: 25866246]

[136] Liu F, Somarowthu S, Pyle AM. Visualizing the secondary and tertiary architectural domains of lncRNA RepA. Nat Chem Biol 2017; 13(3): 282-9.
[http://dx.doi.org/10.1038/nchembio.2272] [PMID: 28068310]

[137] Mercer TR, Mattick JS. Structure and function of long noncoding RNAs in epigenetic regulation. Nat Struct Mol Biol 2013; 20(3): 300-7.
[http://dx.doi.org/10.1038/nsmb.2480] [PMID: 23463315]

[138] Pedram Fatemi R, Salah-Uddin S, Modarresi F, Khoury N, Wahlestedt C, Faghihi MA. Screening for Small-Molecule Modulators of Long Noncoding RNA-Protein Interactions Using AlphaScreen. J Biomol Screen 2015; 20(9): 1132-41.
[http://dx.doi.org/10.1177/1087057115594187] [PMID: 26173710]

[139] Howe JA, Wang H, Fischmann TO, *et al.* Selective small-molecule inhibition of an RNA structural element. Nature 2015; 526(7575): 672-7.
[http://dx.doi.org/10.1038/nature15542] [PMID: 26416753]

[140] Velagapudi SP, Cameron MD, Haga CL, *et al.* Design of a small molecule against an oncogenic noncoding RNA. Proc Natl Acad Sci USA 2016; 113(21): 5898-903.
[http://dx.doi.org/10.1073/pnas.1523975113] [PMID: 27170187]

[141] Liu Y, Luo D, Zhao H, Zhu Z, Hu W, Cheng CH. Inheritable and precise large genomic deletions of non-coding RNA genes in zebrafish using TALENs. PLoS One 2013; 8(10): e76387.
[http://dx.doi.org/10.1371/journal.pone.0076387] [PMID: 24130773]

[142] Gutschner T, Baas M, Diederichs S. Noncoding RNA gene silencing through genomic integration of RNA destabilizing elements using zinc finger nucleases. Genome Res 2011; 21(11): 1944-54.
[http://dx.doi.org/10.1101/gr.122358.111] [PMID: 21844124]

[143] Thakore PI, D'Ippolito AM, Song L, *et al.* Highly specific epigenome editing by CRISPR-Cas9 repressors for silencing of distal regulatory elements. Nat Methods 2015; 12(12): 1143-9.
[http://dx.doi.org/10.1038/nmeth.3630] [PMID: 26501517]

[144] Liu SJ, Horlbeck MA, Cho SW, *et al.* CRISPRi-based genome-scale identification of functional long noncoding RNA loci in human cells. Science 2017; 355(6320): 355.
[http://dx.doi.org/10.1126/science.aah7111] [PMID: 27980086]

[145] Aparicio-Prat E, Arnan C, Sala I, Bosch N, Guigó R, Johnson R. DECKO: Single-oligo, dual-CRISPR deletion of genomic elements including long non-coding RNAs. BMC Genomics 2015; 16: 846.
[http://dx.doi.org/10.1186/s12864-015-2086-z] [PMID: 26493208]

[146] Han J, Zhang J, Chen L, *et al.* Efficient *in vivo* deletion of a large imprinted lncRNA by CRISPR/Cas9. RNA Biol 2014; 11(7): 829-35.
[http://dx.doi.org/10.4161/rna.29624] [PMID: 25137067]

[147] Mizrahi A, Czerniak A, Levy T, *et al.* Development of targeted therapy for ovarian cancer mediated by a plasmid expressing diphtheria toxin under the control of H19 regulatory sequences. J Transl Med 2009; 7: 69.
[http://dx.doi.org/10.1186/1479-5876-7-69] [PMID: 19656414]

[148] Gofrit ON, Benjamin S, Halachmi S, *et al.* DNA based therapy with diphtheria toxin-A BC-819: a phase 2b marker lesion trial in patients with intermediate risk nonmuscle invasive bladder cancer. J Urol 2014; 191(6): 1697-702.
[http://dx.doi.org/10.1016/j.juro.2013.12.011] [PMID: 24342146]

CHAPTER 5

Ros-Mediated Induction of Apoptosis in Breast Cancer Cells

Işıl Yıldırım[*]

Beykent University, Vocational School, Pharmacy Services Program, Biochemistry Area, 34500, Istanbul, Turkey

Abstract: Due to an excessive production of ROS production by the cell's loss of its ability to metabolize, an oxidative stress is expressed. An excessive accumulation of ROS alters not only the biochemical but also genetic and epigenetic mechanisms. These changes could lead to cancer development. In previous studies, it was determined that ROS plays a critical role and the therapy target has also been discussed. So, throughout this manuscript, I aimed to review the role of ROS and its associated signaling pathways in the initiation and progression and apoptosis of breast cancer. This fine distinction of the ROS action depends not only on pro-apoptotic agents and anti-apoptotic agents; ROS leads to the activation of several signaling pathways but also on the duration, type, dosage (concentration) and site of ROS generation and change gene expressions. This study will keep current on the relevant mechanisms of ROS-mediated apoptosis and will guide future studies.

Keywords: ROS, Oxidative Stress, Apoptosis, Capsase, Breast Cancer, MCF7 cells, MDA-MB-231 cells, anticancer drugs, Molecular mechanisms.

INTRODUCTION

Reactive Oxygen Species (ROS) have an important key role in different biological processes. Cell growth, cell growth, cell senescence, the regulation of metabolic mechanisms with signaling pathways, and programmed cell death are examples of the biological processes. Cancer cells proliferation may induce by a higher ROS circumstance or activated signaling pathways that increase the level of intracellular ROS that is not only an expression of several genes but also cause the cancer cells to develop a high metabolic rate. It also has a critical role in the secondary messenger of tumorigenesis and metastasis. Sometimes higher ROS can potentiate cancer cells against oxidative stress-induced cell death. Nitric oxide mediated apoptosis included in upregulation of the tumor suppressor protein p53,

[*] **Corresponding author Işıl Yıldırım:** Beykent University, Vocational School, Pharmacy Services Program, Biochemistry Area, 34500, Istanbul, Turkey; E-mails: isilyld@hotmail.com; Dr.IsilYildirim@hotmail.com

Atta-ur-Rahman (Ed.)

chromatin condensation, DNA laddering, and the activation of proteases known as caspases, associated with changes in the expression of apoptotic united proteins that are related to the Bcl-2 family [1].

In this article, I present and summarize the interaction between ROS status and redox signaling systems, apoptosis and molecular mechanisms associated with it in breast cancer cells (Fig. **1**). This review highlights the elementary value of apoptosis, including its effect on tissue homeostasis and cellular stress.

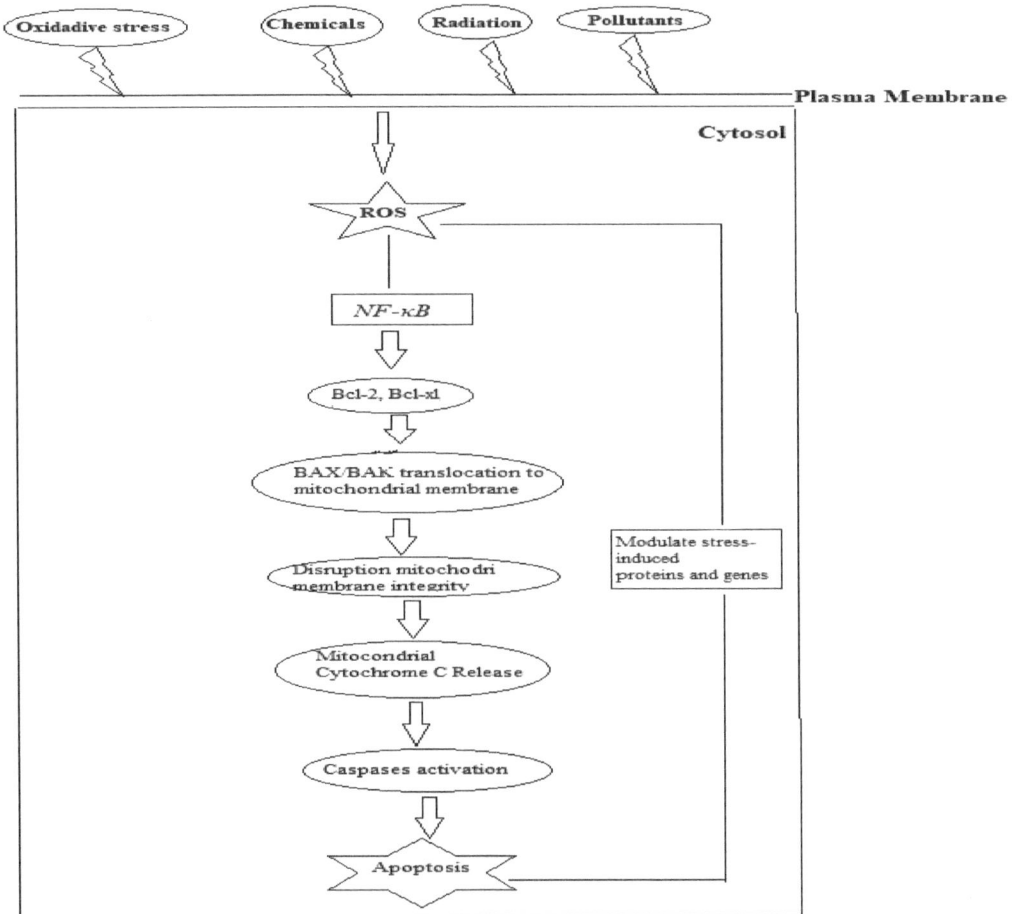

Fig. (1). Proffer mechanism of ROS-induced apoptosis.

THE ROLE OF ROS IN BREAST CANCER CELLS

Breast cancer is a cancer that forms in the cells of the breasts. Many factors that are associated with an increased risk of breast cancer. For example, inherit genes, obesity, malnutrition, radiation exposure, drinking alcohol, drug use, and other factors. The promoted oxidative stress is generally associated with breast cancer formation and progression. Many mechanisms are affected in promoted oxidative stress including genetic variation and ROS generation [2]. For example, in a study to determine ROS mediated cell metastasis in the breast cancer cell lines MCF7 and MDA-MB-231, it was determined that p53-mediated ROS induce p53-phosphorylation by targeting p38MAPK to block NF-kB/p65 nuclear translocation by inhibiting I*k*-Bα-phosphorylation. This inhibition leads to inhibition of metastatic proteins metalloproteinase (MMP)-2 and MMP-9 [3]. It has been reported that LDH-A reduction resulted in an inhibited MCF7 and MDA-MB-231 cancer cell proliferation, high intracellular oxidative stress, and induction of mitochondrial pathway apoptosis [4]. It has been reported that a TGF-β1-ROS-ATM-CREB signaling axis appeared in the macrophage mediated migration of human breast cancer MCF7 cells. In addition, macrophage-conditioned medium leads to the secretion of TGF-β1 in MCF7 cells. This was associated with apoptosis in a fraction of the cells and a generation of ROS and RNS and DNA damage in cells [5]. In Another study, it was determined that Loss of Cdk5, which is located in the mitochondria and is the key marker of apoptosis, activation in breast cancer cells enhanced *via* ROS-mediated cell deaths by targeting the degradation of mitochondrial membranes which leads potential transition pore and mitochondrial fragmentation which is related with an enhanced in both intracellular Ca^{2+} levels and calcineurin activity, and DRP1 S637 de- phosphorylation [6].

It was further reported that Nitric oxide-mediated apoptosis in human breast cancer cells MCF7 and BT-20 occurs with changes in mitochondrial functions and is independent of the activation of the CD95/CD95L pathways [7]. It has been also reported that generate ROS promoted caspase dependent cell death in MCF7 cells by mitochondrial membrane depolarization and cytochrome C release [8, 9]. Another study it has been reported that iodine exhibited cytotoxicity on cultured human breast cancer cell lines, namely MCF7, MDA-MB-231, MDA-MB-453, ZR-75–1, and T-47D. Iodine induced apoptosis in G1 control points in all of the cell lines were tested, except MDA-MB-23. Iodine-induced apoptosis was independent of caspases. Iodine degradation of mitochondrial membrane potential, exhibited antioxidant activity, and can lead to an exhaustion of total cellular thiol content and it showed a diminished Bcl-2 and an up-regulation of Bax in cancer cell [10]. It has been reported that ERK activation was increased in the MDA-MB-231 breast cancer cells treated with triptolide. Tritpolide-induced ERK activation modulated the expression of the Bcl-2 protein family and brought forth caspase-dependent apoptosis. Moreover, upstream of ERK activation lead to generation of ROS and endoplasmic reticulum stress predominantly *via* the PERK-eIF2α pathway [11]. It was Reported that in T47D and MDA-MB-231 breast cancer cells exposed to polychlorinated biphenyls increased in cytotoxic response and ROS formation and PARP-1 activation as being concentration and time-dependent [12].

It has been reported that Icycloheximide promoted apoptotic cell death by increasing the TNF-alpha induction iNOS activity in MCF7 cell [13]. It has also been reported that Tamoxifen-induced apoptosis of MDA-MB-231 BT-20 cancer cells occurred with the activation of caspase-3 and c-Jun NH2-terminal Kinase-1 signaling pathways [14]. Also reported was that IsoobtusilactoneA induces cell cycle arrest and apoptosis through reactive oxygen species/apoptosis signal-regulating kinase 1 signaling pathways in human breast cancer cells [15]. Paclitaxel mediated intracellular ROS generation induced breast cancer deaths [16]. Piperlongumine (PL) inhibited cell viability, increases reactive oxygen species (ROS) and apoptotic cell death in both breast cancer cells and normal cells [17]. It has been identified that Isoliensinine trigger apoptosis in triple-negative human breast cancer cells by targeting ROS generation and p38 MAPK/JNK activation [18]. Also reported was that nardosinen significantly inhibited cell proliferation in a dose-dependent manner in MCF7 breast cancer cells and ınduced apoptosis by caspase pathways and loss of mitochondrial membrane potentials increase of ROS [19].

To determine the role of reactive oxygen species in estrogen dependent breast cancer identified that ROS generated by estrogens affects such as CD1, AKT, NF-κB molecules and Bcl2 anti-apoptotic gene [20]. In a previous study, it was identified that the promoted ROS under low oxygen conditions, HIF-1α transcriptionally was active in breast carcinoma cells [21]. Another study identified that low sodium arsenide mediated ROS formation induced MCF7 epithelial breast cancer cell proliferation and it leads to the activation of NF-κB and an expression of c-Myc protein and an increased in hem oxygenase-1(HO-1) [22]. It has been identified that Estrogen significantly enhanced ROS formation and it lead to increase in heme oxygenase (HO-1), an indicator of oxidative stress levels, when estrogen induced with c-Src inhibitor, Estrogen-induced apoptosis inhibited and blocked Estrogen-induced ROS production [23].

According to a study publish in 2018 year, it was determined that after treatment with metformin which resulted in an increase in Nitric Oxide (NO) production related to apoptosis processes and in the MCF7, CAMA1 and HCC1143 cell lines lead to an enhanced superoxide dismutase activation, but there was no change in the MDA-MB-231 cell line. Moreover, treatment with MTF increased expression catalase which increased in all other cells except CAMA1. Also it was determined that proteins with differentials found between the treated groups and control groups. For example, while the MCF7 cells expression of proteins related to mitochondria and cell cycle diminish, expression of proteins involved in mitochondria and cytoskeleton were enhanced. For MDA-MB-23, MDA-MB-468 and HCC1143 cells there was a diminished the expression of proteins mostly related to mitochondria. Moreover, treatment with metformin lead to diminished activity in the mitochondria B, mRNA processing, ATP binding and DNA replication functional points in all the cell line. Flow cytometry analysis results showed that in the MCF7 and MDAMB231 cells treated with metformin, cell population proportion increased at the G2 / M point relative to the control group, but in the CAMA1 cell, cell populations proportions increased in the G1 phase [24].

Radiation is an important factor in the formation of oxidative stress. Pedram and coworkers in their study identified that in the MCF7 cell, estrogen inhibited UV radiation-induced cytochrome C release, the diminishing of the mitochondrial membrane potential, and apoptotic cell deaths. The same study also determined that UV lead to the formation of a mitochondrial reactive oxygen species (mROS). This is a mROS activation of the c-jun N-terminal kinase (JNK), and protein kinase C (PKC), and Bax of mitochondria. The results of all of these events lead to inducing mitochondrial cell death [25]. It has also been reported that Radio sensitivity promoted by targeting increased Bax expression, reduced Bcl-2 expression, prolonged G2-M arrest, and enhanced apoptosis of MDA-M--231 cell [26].

Another report found that radiation significantly increased oxidative stress and it lead to accumulation of DNA damage and induced apoptosis. Moreover, miR-139-5p is a modulator which causes a radiotherapy response in breast cancer, and its gene expression [27]. Hu and coworkers in their study identified that miR-125b significantly promoted the generation of ROS in MCF7 cells by deoxorubicin treatment and it regulates the drug-resistance of breast cancer cells by targeting HAX-1 [28]. Furthermore, the mitochondria-targeted miR-4485 exhibited a tumor suppressor effect in breast carcinoma cells by targeting negatively regulating mitochondrial RNA processing and mitochondrial functions [29]. Another a study it was identified that an up-regulation of miR-223 increased in TRAIL-induced apoptosis by targeting the mitochondria/ ROS pathway [30]. It has been identified that the expression of miR-15a / miR-16, increased mitochondrial ROS generation, the degradation of mitochondrial membrane potential, followed by Cytochrome-C release into the cytosol that then activated Caspase-3 and Caspase-6/9 leading to intrinsic apoptosis in MCF7, MDAMB-231 breast cancer cells targeted by blocking BMI1 protein expression Additionally, it also inhibits migration [31].

Ionizing radiation can be induced into cancer cells however, these transmissions can also damage neighboring non-cancerous cells. It was also determined that extracellular ROS release induced by gradient radiation was higher than the uniform radiation exposure for 24 and 48 hours in MCF7 cells. Moreover, enhanced apoptosis and a reduction in cell viability gradient radiation stimulation is also increased [32]. Another study showed that ionized radiation induced in radiosensitive parental TNBC cells lead to diminished/ lower levels of ROS and enhanced protein levels of phospho-STAT3 and Bcl-2. Moreover, in a combination with radiation, niclosamide, and a STAT3 inhibitor, treatments resulted in a significant increase of ROS generation and induction of apoptosis both parental and radio resistant TNBC cells [33].

Considering literature information, it is seen that radiation and drugs are effective in ROS formation. ROS mediated oxidative stress has been confirmed by studies that induce apoptosis with different mechanisms.

CONCLUSION

Increased ROS levels, alteration of the redox balance, and the restrictions of redox signaling have an important role and are common hallmarks of cancer progression and resistance to treatment. Because they cause high metabolic activity by influencing in various ways such as cellular signaling, peroxisomal activity, mitochondrial dysfunction, activation of oncogene, and an increased enzymatic activity in cancer cells. Changes in the ROS mediated metabolic process might cause an alteration in genetic and tumor microenvironment either by inducing the formation or death of cancer cells. In this article, I summarized the interaction between the ROS status and apoptosis or molecular mechanisms associated with it in breast cancer cells. When the studies are examined, it is seen that ROS formation in breast cancer is caused by an expression of inflammation-mediated factors and apoptotic proteins. So, performing additional detailed studies on reactive oxygen species will provide a potential benefit in breast cancer treatment.

ACKNOWLEDGMENT

Not applicable

CONFLICT OF INTEREST

I have no any conflict of interest. I have no commercial associations.

REFERENCES

[1]   Brüne B, Sandau K, von Knethen A. Apoptotic cell death and nitric oxide: activating and antagonistic transducing pathways. Biochemistry (Mosc) 1998; 63(7): 817-25.
       [PMID: 9721334]

[2]   Omar MEAS, Eman RY, Hafez HF. The Antioxidant Statues of the Plasma in Patients with Breast Cancer Undergoing Chemotherapy. Open J Mol Integr Physiol 2011; 1: 29-35.
        [http://dx.doi.org/10.4236/ojmip.2011.13005]

[3]   Adhikary A, Mohanty S, Lahiry L, Hossain DM, Chakraborty S, Das T. Theaflavins retard human breast cancer cell migration by inhibiting NF-kappaB *via* p53-ROS cross-talk. FEBS Lett 2010; 584(1): 7-14.
        [http://dx.doi.org/10.1016/j.febslet.2009.10.081] [PMID: 19883646]

[4]   Wang ZY, Loo TY, Shen JG, *et al.* LDH-A silencing suppresses breast cancer tumorigenicity through induction of oxidative stress mediated mitochondrial pathway apoptosis. Breast Cancer Res Treat 2012; 131(3): 791-800.
        [http://dx.doi.org/10.1007/s10549-011-1466-6] [PMID: 21452021]

[5]   Singh R, Shankar BS, Sainis KB. TGF-β1-ROS-ATM-CREB signaling axis in macrophage mediated migration of human breast cancer MCF7 cells. Cell Signal 2014; 26(7): 1604-15.
        [http://dx.doi.org/10.1016/j.cellsig.2014.03.028] [PMID: 24705025]

[6]   NavaneethaKrishnan S, Rosales JL, Lee KY. Loss of Cdk5 in breast cancer cells promotes ROS-mediated cell death through dysregulation of the mitochondrial permeability transition pore. Oncogene 2018; 37(13): 1788-804.
        [http://dx.doi.org/10.1038/s41388-017-0103-1] [PMID: 29348461]

[7]   Umansky V, Ushmorov A, Ratter F, *et al.* Nitric oxide-mediated apoptosis in human breast cancer cells requires changes in mitochondrial functions and is independent of CD95 (APO-1/Fas). Int J Oncol 2000; 16(1): 109-17.
        [PMID: 10601555]

[8]   Weitsman GE, Ravid A, Liberman UA, Koren R, Vitamin D. Vitamin D enhances caspase-dependent and -independent TNFalpha-induced breast cancer cell death: The role of reactive oxygen species and mitochondria. Int J Cancer 2003; 106(2): 178-86.
        [http://dx.doi.org/10.1002/ijc.11202] [PMID: 12800192]

[9]   Weitsman GE, Koren R, Zuck E, Rotem C, Liberman UA, Ravid A. Vitamin D sensitizes breast cancer cells to the action of H2O2: mitochondria as a convergence point in the death pathway. Free Radic Biol Med 2005; 39(2): 266-78.
        [http://dx.doi.org/10.1016/j.freeradbiomed.2005.03.018] [PMID: 15964518]

[10]   Shrivastava A, Tiwari M, Sinha RA, *et al.* Molecular iodine induces caspase-independent apoptosis in human breast carcinoma cells involving the mitochondria-mediated pathway. J Biol Chem 2006; 281(28): 19762-71.
        [http://dx.doi.org/10.1074/jbc.M600746200] [PMID: 16679319]

[11]   Tan BJ, Chiu GNC. Role of Oxidative Stress and ERK Activation in Triploide-induced Apoptosis. Int J Oncol 2013; 42: 1605-12.
        [http://dx.doi.org/10.3892/ijo.2013.1843] [PMID: 23467622]

[12]   Lin CH, Lin PH. Induction of ROS formation, poly(ADP-ribose) polymerase-1 activation, and cell death by PCB126 and PCB153 in human T47D and MDA-MB-231 breast cancer cells. Chem Biol Interact 2006; 162(2): 181-94.
        [http://dx.doi.org/10.1016/j.cbi.2006.06.009] [PMID: 16884709]

[13]   Binder C, Schulz M, Hiddemann W, Oellerich M. Induction of inducible nitric oxide synthase is an essential part of tumor necrosis factor-alpha-induced apoptosis in MCF-7 and other epithelial tumor cells. Lab Invest 1999; 79(12): 1703-12.
        [PMID: 10616218]

[14]   Mandlekar S, Yu R, Tan TH, Kong AN. Activation of caspase-3 and c-Jun NH_2-terminal kinase-1 signaling pathways in tamoxifen-induced apoptosis of human breast cancer cells. Cancer Res 2000; 60(21): 5995-6000.
        [PMID: 11085519]

[15]   Kuo PL, Chen CY, Hsu YL, Isoobtusilactone A. Isoobtusilactone A induces cell cycle arrest and apoptosis through reactive oxygen species/apoptosis signal-regulating kinase 1 signaling pathway in human breast cancer cells. Cancer Res 2007; 67(15): 7406-20.
[http://dx.doi.org/10.1158/0008-5472.CAN-07-1089] [PMID: 17671211]

[16]   Alexandre J, Hu Y, Lu W, Pelicano H, Huang P. Novel action of paclitaxel against cancer cells: bystander effect mediated by reactive oxygen species. Cancer Res 2007; 67(8): 3512-7.
[http://dx.doi.org/10.1158/0008-5472.CAN-06-3914] [PMID: 17440056]

[17]   Raj L, Ide T, Gurkar AU, et al. Selective killing of cancer cells by a small molecule targeting the stress response to ROS. Nature 2011; 475(7355): 231-4.
[http://dx.doi.org/10.1038/nature10167] [PMID: 21753854]

[18]   Zhang X, Wang X, Wu T, et al. Isoliensinine induces apoptosis in triple-negative human breast cancer cells through ROS generation and p38 MAPK/JNK activation. Sci Rep 2015; 5: 12579.
[http://dx.doi.org/10.1038/srep12579] [PMID: 26219228]

[19]   Shahali A, Ghanadian M, Jafari SM, Aghaei M. Mitochondrial and caspase pathways are involved in the induction of apoptosis by nardosinen in MCF-7 breast cancer cell line. Res Pharm Sci 2018; 13(1): 12-21.
[http://dx.doi.org/10.4103/1735-5362.220963] [PMID: 29387107]

[20]   Johar R, Sharma R, Kaur A, Mukherjee TK. Role of Reactive Oxygen Species in Estrogen Dependant Breast Cancer Complication. Anticancer Agents Med Chem 2015; 16(2): 190-9.
[http://dx.doi.org/10.2174/1871520615666150518092315] [PMID: 25980816]

[21]   De Luca A, Sanna F, Sallese M, et al. Methionine sulfoxide reductase A down-regulation in human breast cancer cells results in a more aggressive phenotype. Proc Natl Acad Sci USA 2010; 107(43): 18628-33.
[http://dx.doi.org/10.1073/pnas.1010171107] [PMID: 20937881]

[22]   Ruiz-Ramos R, Lopez-Carrillo L, Rios-Perez AD, De Vizcaya-Ruíz A, Cebrian ME. Sodium arsenite induces ROS generation, DNA oxidative damage, HO-1 and c-Myc proteins, NF-kappaB activation and cell proliferation in human breast cancer MCF-7 cells. Mutat Res 2009; 674(1-2): 109-15.
[http://dx.doi.org/10.1016/j.mrgentox.2008.09.021] [PMID: 18996220]

[23]   Fan P, Griffith OL, Agboke FA, et al. c-Src modulates estrogen-induced stress and apoptosis in estrogen-deprived breast cancer cells. Cancer Res 2013; 73(14): 4510-20.
[http://dx.doi.org/10.1158/0008-5472.CAN-12-4152] [PMID: 23704208]

[24]   Trilla-Fuertes L, Gámez-Pozo A, Arevalillo JM, et al. Molecular characterization of breast cancer cell response to metabolic drugs. Oncotarget 2018; 9(11): 9645-60.
[http://dx.doi.org/10.18632/oncotarget.24047] [PMID: 29515760]

[25]   Pedram A, Razandi M, Wallace DC, Levin ER. Functional estrogen receptors in the mitochondria of breast cancer cells. Mol Biol Cell 2006; 17(5): 2125-37.
[http://dx.doi.org/10.1091/mbc.e05-11-1013] [PMID: 16495339]

[26]   Zhou KX, Xie LH, Peng X, et al. CXCR4 antagonist AMD3100 enhances the response of MDA-M--231 triple-negative breast cancer cells to ionizing radiation. Cancer Lett 2018; 418: 196-203.
[http://dx.doi.org/10.1016/j.canlet.2018.01.009] [PMID: 29317253]

[27]   Pajic M, Froio D, Daly S, et al. miR-139-5p Modulates Radiotherapy Resistance in Breast Cancer by Repressing Multiple Gene Networks of DNA Repair and ROS Defense. 2018. Transl Sci
[http://dx.doi.org/10.1158/0008-5472.CAN-16-3105]

[28]   Hu G, Zhao X, Wang J, et al. miR-125b regulates the drug-resistance of breast cancer cells to doxorubicin by targeting HAX-1. Oncol Lett 2018; 15(2): 1621-9.
[PMID: 29434858]

[29]   Sripada L, Singh K, Lipatova AV, et al. hsa-miR-4485 regulates mitochondrial functions and inhibits the tumorigenicity of breast cancer cells. J Mol Med (Berl) 2017; 95(6): 641-51.

[http://dx.doi.org/10.1007/s00109-017-1517-5] [PMID: 28220193]

[30]   Sun X, Li Y, Zheng M, Zuo W, Zheng W. MicroRNA-223 Increases the Sensitivity of Triple-Negative Breast Cancer Stem Cells to TRAIL-Induced Apoptosis by Targeting HAX-1. PLoS One 2016; 11(9): e0162754.
[http://dx.doi.org/10.1371/journal.pone.0162754] [PMID: 27618431]

[31]   Patel N, Garikapati KR, Ramaiah MJ, Polavarapu KK, Bhadra U, Bhadra MP. miR-15a/miR-16 induces mitochondrial dependent apoptosis in breast cancer cells by suppressing oncogene BMI1. Life Sci 2016; 164: 60-70.
[http://dx.doi.org/10.1016/j.lfs.2016.08.028] [PMID: 27596816]

[32]   Zhang D, Zhou T, He F, *et al.* Reactive oxygen species formation and bystander effects in gradient irradiation on human breast cancer cells. Oncotarget 2016; 7(27): 41622-36.
[http://dx.doi.org/10.18632/oncotarget.9517] [PMID: 27223435]

[33]   Lu L, Dong J, Wang L, *et al.* Activation STAT3 and Bcl-2 and Reduction of Species (ROS) Promote Radioresistence in Breast Cancer and Overcome of Radio Resistance with Niclosamid. Oncogene 2018.
[http://dx.doi.org/10.1038/s41388-018-0340-y]

CHAPTER 6

Generations of Epidermal Growth Factor Receptor Tyrosine Kinase Inhibitors: A Battle against Drug Resistant Lung Cancer

Kenneth K.W. To[*] and Wing-Sum Tong

School of Pharmacy, Faculty of Medicine, The Chinese University of Hong Kong, Hong Kong SAR, China

Abstract: The discovery of epidermal growth factor receptor tyrosine kinase inhibitors (EGFR TKI), such as gefitinib and erlotinib, has produced remarkable clinical response in a sub-population of lung cancer patients harboring the sensitizing EGFR mutations (L858R or exon 19 deletion). However, their successful clinical application is significantly hindered by the development of acquired resistance predominantly caused by the secondary EGFR T790M mutation, usually occurring within a year after the initial TKI therapy. Second generation irreversible EGFR TKIs have been developed to bind covalently to Cys-797 of the EGFR kinase binding domain in order to bypass these EGFR T790M mutations. However, these irreversible EGFR TKIs are not sufficiently effective against the resistant cells *in vivo* at clinically achievable drug concentrations. In order to overcome resistant mutations and also to reduce toxic effects, highly potent third generation EGFR TKIs have been designed against EGFR T790M-bearing cancer while sparing the wild-type receptor. However, acquired resistance to the third generation TKIs has already been reported, which is mediated by the induction of another secondary EGFR C797S mutation and in some instances MET amplification. This chapter summarizes the current research in the development of EGFR TKIs, mainly focusing on pharmacological properties, safety and clinical status.

Keywords: Afatinib, Dacomitinib, Drug Resistance, Epidermal Growth Factor Receptor (EGFR), Gefitinib, Molecular-Targeted Therapy, Mutant Selective EGFR Inhibitor, Osimertinib, Non-Small Cell Lung Cancer, Tyrosine Kinase Inhibitor.

INTRODUCTION

Lung cancer is the leading cause of cancer deaths worldwide [1]. Non-small cell lung cancer (NSCLC) accounts for 85% of all lung cancer cases and it is further

---

[*] **Corresponding author Kenneth K.W. To:** School of Pharmacy, Faculty of Medicine, The Chinese University of Hong Kong, Hong Kong SAR, China; Tel: (852)39438017; E-mail: kennethto@cuhk.edu.hk

Atta-ur-Rahman (Ed.)

histologically classified into three main subtypes: adenocarcinoma, squamous cell carcinoma and large-cell carcinoma [2]. Traditional cytotoxic chemotherapy is the major treatment option for advanced NSCLC, but treatment response rates are modest. Therefore, substantial research efforts have been made to identify novel targets for NSCLC treatment.

EGFR is one of the four members of HER (Human Epidermal growth factor Receptor) family of receptor tyrosine kinases: HER1/EGFR, HER2, HER3 and HER4. These receptor family members are transmembrane proteins that recognize different growth-related ligands, such as epidermal growth factor (EGF) and transforming growth factor alpha (TGFα). Receptor-ligand interaction induces a conformational change, which promotes a homo- or hetero-dimerization between receptors, followed by auto-phosphorylation of the receptors themselves at tyrosine residues. This process results in the activation of intracellular signaling pathways that play a key role in regulating cell growth and survival [3]. In certain types of NSCLC, the *EGFR* gene is mutated and become constitutively active (which occur in exons 18 to 21 encoding the adenosine triphosphate (ATP)-binding pocket of the kinase domain) [4 - 6], leading to "addictive" oncogenic signaling. Common EGFR activating mutations (90%) include deletion at exon 19 (Del 19) and point mutations in exon 21 most commonly resulting in a substitution of arginine for leucine at amino acid 858 (L858R) [4, 7, 8]. EGFR tyrosine kinase inhibitors (TKIs) were specifically designed to target these "addictive" oncogenic signaling and disrupt the associated downstream pathways to selectively kill cancer cells.

FIRST GENERATION OF EGFR-TKIS

First generation of EGFR-TKIs include gefitinib (ZD1839, Iressa®), erlotinib (OSI-774, Tarceva®) and icotinib (BPI-2009, Conmana™).

Gefitinib was initially approved by FDA on 5 May 2003 as a third-line treatment of patients with locally advanced or metastatic NSCLC after failure of both platinum-based and docetaxel chemotherapies, but its approval was withdrawn in June 2005 for use in new patients (without genetic screening) due to a lack of evidence of prolonged survival. However, on 13 July 2015, gefitinib was approved by FDA again as a first-line treatment of patients with metastatic NSCLC whose tumors have EGFR Del 19 or L858R mutations (Table **1**).

Erlotinib was first approved by FDA on 18 November 2004 for the treatment of locally advanced or metastatic NSCLC that has failed at least one prior chemotherapy regimen. It was further approved on 16 April 2010 for the maintenance treatment of patients with locally advanced or metastatic NSCLC whose disease has not progressed after four cycles of platinum-based first-line chemotherapy, and on 18 October 2016 for the first-line treatment of patients with metastatic NSCLC whose tumors have EGFR Del 19 or L858R mutations. However, on 18 October 2016, its first and second indications were restricted only to metastatic NSCLC patients bearing the EGFR Del 19 or L858R mutations (Table **1**).

Table 1. Current status of different generations of EGFR-TKIs monotherapy.

EGFR-TKIs	Type of TKI	Current status	Indications
First-generation EGFR-TKIs			
Gefitinib	Reversible	First-time approved by FDA (5 May 2003)	Third-line treatment of patients with locally advanced or metastatic NSCLC after failure of both platinum-based and docetaxel chemotherapies
		Withdrawn by FDA (June 2005)	Withdrawal for use in new patients due to a lack of evidence that it extended life
		Second-time approved by FDA (13 July 2015)	First-line treatment of patients with metastatic NSCLC whose tumors have EGFR Del 19 or L858R mutations
Erlotinib		Approved by FDA (First indication, 18 November 2004)	Treatment of patients with locally advanced or metastatic NSCLC after failure of at least one prior chemotherapy regimen
		Approved by FDA (Second indication, 16 April 2010)	Maintenance treatment of patients with locally advanced or metastatic NSCLC whose disease has not progressed after four cycles of platinum-based first-line chemotherapy
		Approved by FDA (Third indication, 14 May 2013)	First-line treatment of patients with metastatic NSCLC whose tumors have EGFR Del 19 or L858R mutations
		Indications amended (18 October 2016)	Limit the first and second indication to patients with metastatic NSCLC whose tumors have EGFR Del 19 or L858R mutations
Icotinib		Only approved by CFDA (June 2011)	Treatment of patients with locally advanced or metastatic NSCLC after failure of at least one prior chemotherapy regimen
Second-generation EGFR-TKIs			
Afatinib	Irreversible	Approved by FDA (11 July 2013)	Treatment of patients with locally advanced or metastatic NSCLC with EGFR mutation(s)
Dacomitinib		Phase III clinical trial ongoing	Not applicable
Canertinib		No longer investigated	
Neratinib		No longer investigated	

(Table 1) cont.....

EGFR-TKIs	Type of TKI	Current status	Indications
Third-generation EGFR-TKIs			
Osimertinib	Irreversible	Approved by FDA (10 November 2015)	Treatment of advanced EGFR T790M-positive NSCLC
Rociletinib		All clinical trials halted in May 2016	Not applicable
Olmutinib		Only approved by South Korea (17 May 2016)	Second-line treatment of patients with locally advanced or metastatic EGFR T790M mutation-positive NSCLC, who had been previously treated with an EGFR-TKI
ASP8273		Phase I, II, III clinical trials ongoing (trials on ASP8273 discontinued in May 2017)	Not applicable
EGF816			
PF-06747775			

Icotinib was developed in China and only approved by the China Food and Drug Administration (CFDA) in June 2011 for the treatment of locally advanced or metastatic NSCLC in patients who failed at least one prior chemotherapy (Table 1). Notably, it costs only 70% of gefitinib and 60% of erlotinib.

STRUCTURES AND MECHANISM OF ACTION

The structures of gefitinib, erlotinib and icotinib, all of which are synthetic small molecule EGFR TKIs bearing an anilinoquinazoline backbone, are shown in Fig. (1). They compete with ATP for binding to the tyrosine kinase domain of EGFR, thus inhibiting auto-phosphorylation of receptors at tyrosine residues and initiation of downstream signaling cascades to exert their anticancer activity. Of note, all of these EGFR inhibitors are administered orally.

GEFITINIB

The optimal biological dose is 250 mg/day. A randomized, double-blind phase II IDEAL (Iressa Dose Evaluation in Advanced Lung Cancer) 1 trial was conducted to compare the clinical efficacy and tolerability between gefitinib 250 mg/day and 500 mg/day in patients with advanced NSCLC [9]. The lower dose was adopted in the clinic because it showed the same efficacy but less severe adverse events compared to the higher dose.

Clinical Efficacy

Gefitinib was demonstrated to be superior to standard chemotherapy as a first-line treatment for advanced NSCLC whose tumors harbor EGFR Del 19 or L858R mutations based on the results of a key phase III clinical trial, IPASS, (Iressa Pan-Asia Study) conducted in clinically selected patients in East Asia who had advanced NSCLC [10]. In a subgroup of 261 patients who were positive for EGFR mutations, gefitinib group compared to chemotherapy group (carboplatin and paclitaxel) showed significantly improved objective response rate (71.2% *versus* 47.3%) and longer progression-free survival (10.9 months *versus* 7.4 months). An open label, phase III trial (WJTOG3405) conducted in Japan in chemotherapy-naïve patients with stage IIIB/IV NSCLC or postoperative recurrence harboring EGFR mutations (either exon 19 deletion of L858R point mutation also revealed that the gefitinib group gave rise to significantly longer progression-free survival than the chemotherapy (cisplatin plus docetaxel) group (9.2 months *versus* 6.3 months, respectively, HR0.489, p < 0.0001) [11]. Similarly, another randomized, multicenter, phase III trial also showed that the use of gefitinib resulted in progression-free survival that is twice as long as that obtained with the use of carboplatin plus paclitaxel in patients with mutated-EGFR NSCLC (10.8 *versus* 5.4 months, respectively; HR 0.30, 95% CI 0.22-0.41, p < 0.001) [6]. Other key phase III studies on gefitinib are summarized in Table (**2**).

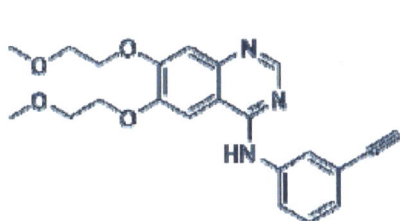

Gefitinib

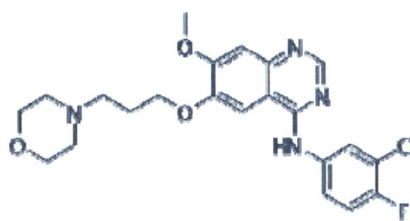

Erlotinib

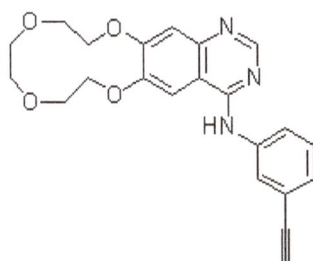

Icotinib

Fig. (1). Chemical structures of first-generation EGFR TKIs.

Table 2. Key Phase III clinical trials of gefitinib monotherapy.

Trial	treatment	No. of patients	Selected population	ORR (%)	PFS (months)	OS (months)	Reference
INTEREST	Gefitinib	723	Advanced NSCLC patients pretreated with ≥ one platinum-based regimen	9.1 (p=0.3257)	2.2 (HR HR 1.04, 95% CI: 0.93-1.18)	7.6 (HR 1.02, 95% CI: 0.91-1.15)	[99]
	Docetaxel	710		7.6	2.7	8.0	
V-15-32	Gefitinib	245	Advanced NSCLC patients pretreated with chemotherapy		2.0 (HR 0.90, p=0.335)	Noninferiority in OS was not achieved	[100]
	Docetaxel	244			2.0		
IPASS	Gefitinib	609	Patients with stage IIIB or IV NSCLC who have not received any previous chemotherapy		All patients: 5.7 (HR 0.74, p<0.001) EGFR mutation positive: 8.5 (HR 0.48, p<0.001)		[10]
	Carboplatin/paclitaxel	608			All patients: 5.8 EGFR mutation positive: 6.0		
ISTANA	Gefitinib	82	Patients with stage IIIB or IV NSCLC who have received only one previous platinum-doublet chemotherapy	28.1 (p=0.0007)	3.3 (HR 0.729, p=0.0441)	14.1 (HR 0.870, p = 0.437)	[101]
	Docetaxel	79		7.6	3.4	12.2	
WJTOG 0203	Gefitinib	298	Patients with stage IIIB or IV NSCLC who have received not received any chemotherapy		4.3 (HR 0.68, p < 0.001)	12.9 (HR 0.86, p = 0.11)	[102]
	Platinum-doublet	300			4.6	13.7	
EORTC 08021/ILCP 01/03	Gefitinib	86	NSCLC patients not progressing after four cycles of platinum-based chemotherapy		4.1 (HR 0.61, p = 0.0015)	10.9 (HR 0.83, p = 0.2)	[103]
	Placebo	87			2.9	9.4	
First-SIGNAL	Gefitinib	159	Patients with stage IIIB or IV NSCLC who have received not received any chemotherapy		All patients: 5.8 (HR 1.198, p = 0.138) EGFR mutation positive: 8.0 (HR 0.544, 95% CI: 0.269-1.100)	22.3 (HR 0.932, p = 0.604)	[104]
	Gemcitabine & cisplatin	154			All patients: 6.4 EGFR mutation positive: 6.3	22.9	

(Table 2) cont.....

Trial	treatment	No. of patients	Selected population	ORR (%)	PFS (months)	OS (months)	Reference
NEJ002	Gefitinib	114	Chemo-naïve NSCLC patients with sensitive EGFR gene mutations		10.8 (HR 0.322; 95% CI 0.236-0.438, p < 0.001)	27.7 (HR 0.887; 95% CI 0.634-1.241, p =0.483)	[105]
	Carboplatin & paclitaxel	114			5.4	26.6	
INFORM	Gefitinib	148	Locally advanced/metastatic NSCLC patients who had achieved disease control after first-line platinum-based doublet chemotherapy			Total population: 18.97 (HR 0.88, p = 0.335) EGFR mutation positive: 46.87 (HR 0.39, p = 0.036)	[106]
	Placebo	148				Total population: 16.00 EGFR mutation positive: 20.97	
WJOG 5108L	Gefitinib	280	Patients with stage IIIB or IV NSCLC who have received at least one chemotherapy regimen		6.5 (HR 1.125, 95% CI: 0.940-1.347)	22.8 (HR 1.038, 95% CI: 0.833-1.294)	[107]
	Erlotinib	281			7.5	24.5	
Indian trial (CTRI/2015/08/006113)	Gefitinib	145	NSCLC patients with EGFR-activating mutation-positive stage IIIB or stage IV adenocarcinoma in the first-line setting	63.5% (p=0.003)	8.4 (HR 95% CI: 0.513-0.851; p=0.001)		[108]
	Pemetrexed & carboplatin	145		45.3%	5.6		

[a] weeks

Safety and Tolerability

Gefitinib was well-tolerated. The maximum tolerated dose (MTD) of gefitinib is 800 mg [12]. The most common drug-related adverse events were skin rash or acne (66.2%) and diarrhea (46.6%). Few grade 3 or above adverse events occurred (< 4%) [5]. Toxicity and dose reduction recommendation for gefitinib and other EGFR TKIs that will be discussed later is summarized in Table **3**.

Quality of Life (QoL) Analysis

In a phase 3 study (North East Japan Study Group 002 trial), the QoL of chemotherapy-naïve NSCLC patients with sensitive EGFR-mutated and advanced tumor after treatment with gefitinib and chemotherapy (carboplatin and paclitaxel) were compared [6, 13]. Time to deterioration from baseline on physical, mental and life well-being QoL scales were assessed. These parameters significantly favored gefitinib over the chemotherapy arm, thus supporting the clinical application of gefitinib as the standard first-line treatment for advanced EGFR-mutated NSCLC.

Table 3. Toxicity and dose reduction recommendation for EGFR TKIs.

EGFR TKI	Grade 3 – 4 Adverse Events	Dose Reduction Recommendation	Reference
First-generation			
Gefitinib	Hepatotoxicity (13.0%) Rash (2.2%) Diarrhea (2.2%)	Change the everyday dosing schedule to every 2 days schedule	[92,93]
Erlotinib	Rash (18.1%) Diarrhea (3.3%)	Reduce from the standard dose of 150 mg/day to 100 mg/day	[92,94]
Icotinib	Available data for both efficacy and safety of icotinib are limited. Rash (1%), pain (2%), raised aminotransferase (1%)	No patient needed dose reduction in reported clinical trial	[95]
Second-generation			
Afatinib	Diarrhea (12.5%) Rash (9.4%)	Reduce dose by 10 mg/day from the initial dose of 40 mg/day	[92,96]
Dacomitinib	Dermatitis acneiform (14%) Diarrhea (8%) Paronychia (7%) Rash (4%) Hypokalaemia (1%)	In the clinical trial, 38.3% patients required a dose reduction from 45 mg to 30 mg and 27.8% patients required a further reduction to 15 mg	[51]
Canertinib	High dose (150 mg/day): Diarrhea (20.0%) Asthenia (12.9%) Mucositis (5.7%) Rash (4.3%)	Not available	[97]
Neratinib	High dose (320 mg/day): Diarrhea (46%) Vomiting (8%) Dyspnea (8%) Lymphopenia (8%)	Reduce from the maximum tolerated dose of 320 mg/day to 240 mg/day	[59]
Third-generation			
Osimertinib	Highest dose evaluated (240 mg/day): Diarrhea (5%)	No dose-limiting toxicities; Maximum tolerated dose not defined	[64]

(Table 3) cont.....

EGFR TKI	Grade 3 – 4 Adverse Events	Dose Reduction Recommendation	Reference
Rociletinib	Therapeutic dose: Hyperglycemia (22%) QT prolongation (5%) Nausea (2%)	Not established	[78]
Olmutinib	Rash (5%) Palmar-plantar erythrodysesthesia syndrome (4%) Skin toxicity (epidermolysis)	Not established	[98]

ERLOTINIB

The maximal tolerated dose of erlotinib is 150 mg/day, according to the findings of a phase I trial that assessed the feasibility of erlotinib on a protracted, continuous daily schedule [14].

Clinical Efficacy as a First-line Treatment

Erlotinib was demonstrated to be clinically beneficial in objective response rate and progression-free survival as a first-line treatment for advanced NSCLC patients with EGFR Del 19 or L858R mutations. A phase III EURTAC trial (EURopean TArceva *versus* Chemotherapy) randomly assigned 174 patients (1:1) with advanced NSCLC whose tumors harbored EGFR Del 19 or L858R mutations to receive either erlotinib 150 mg/day or chemotherapy (cisplatin plus docetaxel or gemcitabine) [15]. The objective response rate (complete response + partial response) was 58% in erlotinib group and 15% in chemotherapy group. The median progression-free survival was 9.7 months in erlotinib group *versus* 5.2 months in chemotherapy group. However, the OS did not differ significantly between treatment groups. The OPTIMAL study is another important open-label, randomized, phase III trial conducted at 22 centers in China to compare the efficacy and tolerability of erlotinib *versus* standard chemotherapy (gemcitabine plus carboplatin) in the first-line treatment of patients with advanced EGFR mutation-positive NSCLC [16]. Erlotinib was found to confer a significantly longer progression-free survival than standard chemotherapy (13.1 *versus* 4.6 months, respectively; HR 0.16, 95% CI 0.10-0.26; p < 0.0001). Other key phase III trials evaluating erlotinib are summarized in Table **4**.

Clinical Efficacy as a Maintenance Treatment

Erlotinib was initially demonstrated to be clinically beneficial as a maintenance therapy in patients who do not progress after four cycles of chemotherapy by a phase III trial, SATURN (Sequential TArceva in UnResectable NSCLC) [17]; however, another phase III trial, IUNO study, reported that maintenance treatment

Table 4. Key Phase III clinical trials of erlotinib monotherapy.

Trial	Treatment	No. of Patients	Selected Population	ORR (%)	DCR (%)	PFS (months)	OS (months)	Reference
First-line treatment								
OPTIMAL	Erlotinib	83	Patients with histologically confirmed stage IIIB or IV NSCLC and a confirmed activating mutation of EGFR (exon 19 deletion or exon 21 L858R point mutation)			13.1 (HR 0.16, p < 0.0001)		[16]
	Gemcitabine plus carboplatin	82				4.6		
EURTAC	Erlotinib	86	Patients with advanced NSCLC whose tumors harbored EGFR Del 19 or L858R	58		9.7 (HR 0.30, p < 0.001)	19.3 (HR 1.04, p = 0.87)	[15]
	chemotherapy	87		15		5.2	19.5	
ENSURE	Erlotinib	110	Patient with advanced EGFR mutation-positive			11.0 (HR 0.42, p = 0.0001)	26.3 (HR 0.91, p = 0.607)	[109]
	Gemcitabine/cisplatin	107				5.5	25.5	
Second- or third-line treatment								
DELTA	Erlotinib	150	Patients with measurable unresectable stage IIIB and IV NSCLC with at least one or two previous chemotherapy treatment			2 (HR 1.22, p = 0.09)	14.8 (HR 0.91, p = 0.53)	[110]
	Docetaxel	150				3.2	12.2	
Maintenance treatment								
SATURN								[17]
Whole population	Erlotinib	437	Patients with advanced who did not progress after 4 cycles of platinum-based chemotherapy	11.9	40.8	12.3 w[a] (HR 0.7, p<0.0001)	12.0 (HR 0.81, p=0.0088)	
	Placebo	447		5.4	27.4	11.1 w	11.0	
EGFR-T790M	Erlotinib	307				12.3 w		
	Placebo	311				11.1 w		

(Table 4) cont.....

Trial	Treatment	No. of Patients	Selected Population	ORR (%)	DCR (%)	PFS (months)	OS (months)	Reference
IUNO	Early Erlotinib	322	Patients with advanced NSCLC whose tumors do not harbor known EGFR-activing mutations	6.5	61.2	13.0 (HR 0.94, p = 0.48)	9.7 (HR 1.02, p = 0.82)	[18]
	Late erlotinib (placebo)	321		3.7	59.2	12.0	9.5	

ᵃ weeks

with erlotinib had no clinical benefits in patients with advanced NSCLC without known EGFR-activating mutations [18]. SATURN was a randomized, double-blind, multinational trial that involved 889 patients with locally advanced or metastatic NSCLC whose disease did not progress during first-line platinum-based chemotherapy. The patients were randomly assigned (1:1) to receive either erlotinib 150 mg/day or placebo until disease progression or unacceptable toxicity. Results showed that erlotinib treatment after standard chemotherapy improved progression-free survival compared to placebo in both the overall population (12.3 *versus* 11.1 weeks) and in a subgroup of patients with EGFR positive immunohistochemistry (12.3 *versus* 11.1 weeks). Nevertheless, IUNO study showed that maintenance treatment with erlotinib in patients with advanced NSCLC without known EGFR-activating mutations was unfavorable. This trial assigned 643 patients (1:1) with advanced NSCLC whose tumors harbored no known EGFR-activating mutations to receive either maintenance erlotinib ('early erlotinib') or placebo. Patients who progressed on placebo received open-label erlotinib ('late erlotinib'). Disappointingly, no objective response, disease control rate, progression-free survival or overall survival benefits were observed for 'early erlotinib' *versus* second-line 'late erlotinib'. Hence, erlotinib was suggested to be a maintenance therapy only for NSCLC patients whose tumors harbor EGFR-activating mutations (Del 19 or L858R) (Table **4**). To this end, in a large prospective biomarker study [19], patients with the activating EGFR mutations were found to have the greatest progression-free survival benefit from erlotinib maintenance therapy relative to placebo (HR 0.10; p < 0.001). On the other hand, KRAS mutation status was a significant negative prognostic factor for progression-free survival from erlotinib therapy [19].

Quality of Life (QoL) Analysis

A double-blind phase III trial was conducted in NSCLC patients who had progressed after prior chemotherapy [20]. QoL was assessed by European Organization for Research and Treatment of Cancer QLQ-C30 and the lung cancer module QLQ-LC13. The primary endpoints were the time to deterioration of three common lung cancer symptoms: cough, dyspnea, and pain. Erlotinib was found to significantly improve survival and the important aspects of QOL. In another phase III, randomized, open-label study comparing first-line erlotinib *versus* chemotherapy treatment (carboplatin plus gemcitabine) in Chinese patients with advanced EGFR mutation-positive NSCLC (OPTIMAL; CTONG-0802), QoL analyses were also conducted [21]. The QoL and lung cancer symptoms were assessed by using the Functional Assessment of Cancer Therapy-Lung (FACT-L) questionnaire, which assesses the physical well-being, social well-being, emotional well-being and functional well-being of the patients. FACT-L also includes the Lung Cancer Subscale (LCS), which examines seven key lung cancer symptoms including dyspnea, cough, thoracic pain, breathing difficulties, anorexia, weight loss and cognition, and the 21-item Trial Outcome Index (TOI). Patients receiving erlotinib exhibited clinically relevant improvements in QoL compared with the chemotherapy group in total FACT-L, TOI and LCS (P <0.0001 for all scales) [21].

Safety and Tolerability

The most frequent adverse events are skin rash (80%), diarrhea (57%) and fatigue (57%). Grade 3 or above adverse events occurred in 45% patients, with skin rash and diarrhea being the most frequent ones [6].

ICOTINIB

The optimized dose of icotinib is 125 mg three times daily (t.i.d.), according to two open-label phase I clinical trials in patients with stage III/IV NSCLC and failure in prior chemotherapy [22, 23]. Maximum tolerated dose was not reached.

Clinical Efficacy

Icotinib was found to be non-inferior to gefitinib in the treatment for pretreated patients with advanced NSCLC. A randomized, double-blind phase III ICOGEN (ICOtinib *versus* GEfitinib in previously treated NSCLC) trial was conducted in China to compare the efficacy of icotinib and gefitinib [24]. This trial recruited 399 patients with advanced NSCLC who failed at least one platinum-based chemotherapy regimen. The patients were randomly assigned (1:1) to receive either icotinib 125 mg t.i.d. or gefitinib 250 mg/day. The results demonstrated that icotinib is not inferior to gefitinib with respective to objective response rate (27.6% *versus* 27.2%), disease control rate (75.4% *versus* 74.9%), progression free survival (4.6 *versus* 3.4 months) and overall survival (13.3 *vs* 13.9 months) outcomes. Icotinib also showed similar clinical efficacy as gefitinib in patients harboring EGFR mutations. In a subgroup of EGFR-mutant population, the objective response rate was 62.1% in icotinib group and 53.8% in gefitinib group;

the disease control rate was 86.2% in icotinib group and 94.9% in gefitinib group (Table **5**).

Table 5. Key Phase III clinical trials of icotinib (ICOGEN and CONVINCE).

	Icotinib group (n=199)	Gefitinib group (n=196)	Reference
ICOGEN trial			
Whole population			
Objective response rate (%)	27.6	27.2	[24]
Disease control rate (%)	75.4	74.9	
Progression-free survival (months)	4.6 (HR 0.84, p = 0.13)	3.4	
Overall survival (months)	13.3 (HR 1.02, p = 0.57)	13.9	
EGFR-mutant population			
Objective response rate (%)	62.1	53.8	
Disease control rate (%)	86.2	94.9	
Drug-related adverse events (%)			
Skin rash	40	49	
diarrhea	19	28	
CONVINCE trial			
	Icotinib (n = 148)	**Cisplatin/Pemetrexed (n = 148)**	[25]
Whole population			
Progression-free survival (months)	11.2 (HR 0.61, p = 0.006)	7.9	
Overall survival (months)	30.5 (log rank p = 0.885)	32.1	
Patients harboring EGFR exon 19 deletion			
Progression-free survival (months)	11.2 (HR 0.66, p = 0.136)	8.0	
Overall survival (months)	32.3 (log rank p = 0.4066)	38.8	
Patients harboring EGFR exon 21 L858R point mutation			

(Table 5) cont.....

	Icotinib group (n=199)	Gefitinib group (n=196)	Reference
Progression-free survival (months)	11.1 (HR 0.76, p = 0.331)	7.8	
Overall survival (months)	29.1 (log rank p = 0.5259)	26.7	
Treatment-Related Grade 3 or 4 Adverse Events			
%	9.5 (p = 0.001)	24.8	

a phase III, open-label, randomized trial (CONVINCE) conducted in 18 sites in China, the efficacy and safety of first-line icotinib *versus* cisplatin/pemetrexed maintenance therapy were compared in NSCLC patients with EGFR mutation [25]. The progression-free survival was found to be significantly longer in the icotinib group than in the chemotherapy group (11.2 *versus* 7.9 months; hazard ratio, 0.61) [25]. However, no significant difference in overall survival was observed between the icotinib group and the chemotherapy group in the overall population or in the EGFR-mutated subgroups (exon 19 del/exon 21 L858R).

Safety and Tolerability

Icotinib 125 mg t.i.d. was well-tolerated. The most common drug-related adverse events were skin rash (40%) and diarrhea (19%). Most adverse events were mild to moderate (grade 1 or 2) and reversible on continued treatment. Only a few grade 3 or 4 adverse events occurred (<1%). No patient was withdrawn from the ICOGEN trial due to adverse events [24].

Quality of Life (QoL) Analysis

QoL was assessed as a secondary endpoint in a single-arm, multi-center, prospective trial (NCT02486354) studying icotinib in NSCLC patients previously treated with platinum-based chemotherapy [26]. QoL was assessed by the 4th edition of the Functional Assessment of Cancer Therapy-Lung (FACT-L) questionnaires and Lung Cancer Symptoms Subscale (LCS). The study confirmed that icotinib is non-inferior to gefitinib when compared with a previous phase III trial (ICOGEN) [24] in terms of QoL assessment.

EGFR EXON 19 DELETION IS MORE SENSITIVE THAN EXON 21 L858R TO FIRST GENERATION EGFR TKIS

Results from several clinical trials on gefitinib and erlotinib revealed that NSCLC patients with exon 19 deletion had superior response rates, longer progression-free survival and overall survival than patients with L858R [27, 28]. The observation can be explained by the differential effect of different EGFR mutations on the structure of the tyrosine kinase domain and thus the constitutive activation of the kinase and sensitivity to TKIs. The tyrosine kinase domain of EGFR has a N-lobe (smaller) and a C-lobe (larger). Appropriate 3-D arrangement of a C-helix (within the N-lobe) and the activation loop (within the C-lobe) is needed for activation of EGFR tyrosine kinase. Based on molecular modeling and crystallographic studies, EGFR L858R causes the activation loop to "flip out", thus destabilizing the inactive conformation and favoring the active conformation [29]. On the other hand, exon 19 deletion occur in a β3 protein strain adjacent to the C-helix and the shortening of this strand is believed to favor the active conformation of the kinase. It was found that the structural effect from exon 19 deletion may have a greater effect on activation of EGFR tyrosine kinase than L858R.

To this end, some EGFR TKIs may have greater inhibitory effect on a specific mutation than others. This will have to do with the relative affinity of the specific mutant to ATP *versus* the TKI. For example, kinetic analysis of the interaction between EGFR L858R and EGFR exon 19 deletion showed that both mutants have a higher K_M for ATP but a lower K_i for erlotinib, relative to the wild type EGFR receptor [30]. Therefore, the increased sensitivity of a TKI to a specific mutant may be explained by the lower affinity of that mutant to ATP relative to other mutants.

DRUG RESISTANCE TO FIRST GENERATION EGFR TKIS

As discussed above, the discovery of the first generation EGFR TKIs has led to unprecedented clinical response in the subset of lung cancer patients carrying the sensitizing EGFR mutations [4]. However, their usefulness is severely compromised by drug resistance mediated by various mechanisms [31]. Some patients do not respond to the drugs due to primary resistance, and all patients with initial response eventually relapse due to acquired resistance.

The major mechanisms contributing to drug resistance to first generation EGFR TKIs are summarized in Fig. (**2**). Primary resistance is the intrinsic unresponsiveness to the drug. A key mechanism leading to intrinsic resistance is the presence of a non-sensitive EGFR mutation (usually insertion mutation in exon 20). In the presence of exon 20 insertions, additional amino acid residues are added at the N-lobe of EGFR (M766 to C775), which controls ATP and EGFR-TKI binding. The insertions favor the kinase domain to adopt an active conformation [32]. Moreover, most exon 20 insertion mutations were found to reduce affinity of EGFR TKIs to the kinase domain [33]. The gatekeeper T790M point mutation, though rarely found in tumors before drug exposure to EGFR

TKIs, has also been identified as a possible cause of intrinsic resistance [33]. Another distinct EGFR mutation featured by an in-frame deletion of exons 2-7 in the extracellular domain of EGFR, commonly known as variant III (EGFRvIII), is constitutively active and thereby causing intrinsic resistance [34]. Other mechanisms include KRAS mutation and PTEN loss/deficiency. KRAS mutations occur in about 30% of lung adenocarcinomas and 5% of squamous carcinomas. They are related to a history of tobacco use. Since the majority of patients with KRAS mutations have wild type EGFR, they do not respond to EGFR-TKIs. On the other hand, PTEN loss/deficiency allows the activation of Akt independent of TK receptor status, and thus resulting in hyperactivity of the PI3K-Akt pathway to drive cancer survival [35]. The proapoptotic Bcl-2 family member protein BIM is a critical mediator of EGFR TKI-induced apoptosis in EGFR mutant NSCLC. Therefore, cancer cells bearing a deletion polymorphism of BIM mRNA are able to evade the drug-induced apoptosis, thereby becoming intrinsically resistant [36].

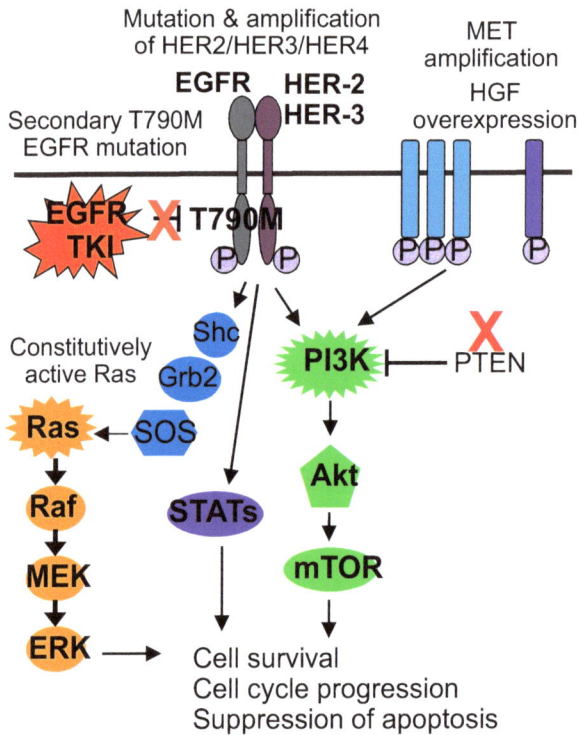

Fig. (2). Mechanisms of drug resistance to first-generation EGFR TKIs.

Acquired resistance to EGFR TKIs usually develops over a period of 12 months upon drug treatment. The most prominent mechanism, which occurs in about 50% NSCLC patients with resistance to EGFR-TKIs, is EGFR T790M mutation-a substitution of threonine with a bulkier residue, methionine, at position 790 at exon 20 of EGFR. The EGFR T790M mutation mediates resistance by (i) sterically hindering EGFR-TKIs from binding to EGFR-TKIs, and (ii) enhancing the affinity of the EGFR kinase for ATP over EGFR-TKIs. On the other hand, aberrant regulation of parallel oncogenic pathways (such as MET amplification and HER2 mutation), which bypass the inhibition of EGFR signaling, has also been reported in resistant cancers. Furthermore, phenotypic transformation, such as epithelial to mesenchymal transition and transformation into other lung cancer histological subtypes such as small cell cancer, have also been reported to contribute to acquired EGFR-TKI resistance [35].

DRUG COMBINATION APPROACHES TO OVERCOME RESISTANCE TO FIRST GENERATION EGFR TKIS

A comprehensive review about the drug combination approaches to overcome resistance to first generation EGFR TKIs can be found in one of our recent review articles [37]. Clinically approved drugs and/or other investigational agents have been combined with EGFR TKIs to overcome the various mechanism of drug resistance. This approach aims at circumventing drug resistance mediated by the so-called bypass signaling mechanism by targeting horizontal or vertical oncogenic signaling pathways or both. The rational combinations of different molecular targeted drugs to inhibit multiple signaling pathways simultaneously have been studied. Besides, the repositioning of drugs with indications other than oncology to combine with EGFR TKIs has also emerged as a promising approach.

SECOND GENERATION OF EGFR-TKIS

Second generation of EGFR-TKIs include afatinib (BIBW2992, Giotrif®, Gilotrif®), dacomitinib (PF-00299804, PF-299), canertinib (CI-1033) and neratinib (HKI-272). Among them, only afatinib has received FDA approval on 11 July 2013 for the treatment of patients with locally advanced or metastatic NSCLC with EGFR mutation(s). Dacomitinib is still undergoing phase III clinical trials, whereas canertinib and neratinib are no longer pursued as a treatment for advanced NSCLC due to their discouraging clinical findings (Table **1**).

STRUCTURE AND MECHANISM OF ACTION

The structures of afatinib, dacomitinib, canertinib and neratinib, all of which are derivatives of anilinoquinazoline compounds, are shown in Fig. (**3**). They are all orally administered. Unlike the first-generation EGFR TKIs, they act as irreversible pan-inhibitor (HER1/EGFR, HER2 and/or HER 4) for the HER family of tyrosine kinases. They bind covalently to cysteine residues within the catalytic domain of HER family (Cys773 of EGFR, Cys805 of HER2 and Cys803 of HER4). This prevents dimerization of EGFR with HER2 or HER4, thus inhibiting HER signaling. However, they also inhibit wild-type EGFR in normal tissues, so their therapeutic benefits were hindered by toxicities, such as diarrhea and skin rash.

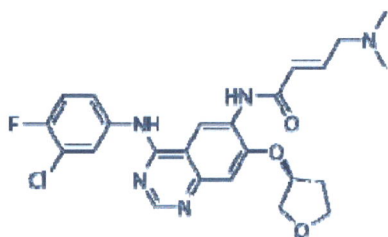

Afatinib

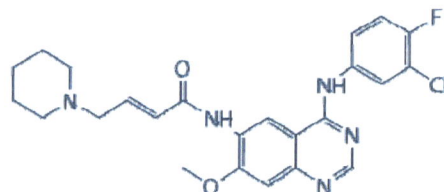

Dacomitinib

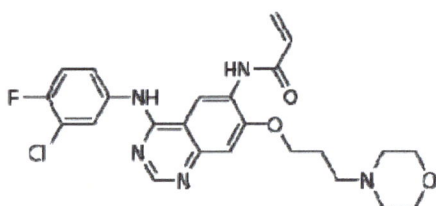

Canertinib

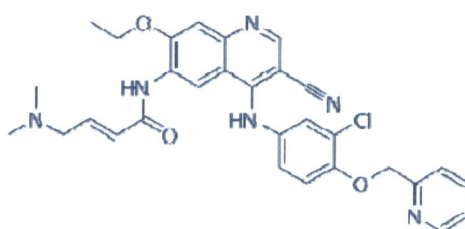

Neratinib

Fig. (3). Chemical structures of second-generation EGFR TKIs.

AFATINIB

The optimized dose of afatinib was 40 mg/day. LUX-Lung 2, a multicenter phase II, open-label trial, evaluated the activity of afatinib at 40 and 50 mg/day as a first or second-line treatment in EGFR-TKI-naïve patients with EGFR-mutated advanced NSCLC [38]. The primary endpoint was ORR. Patients showed no differences in ORR between two dose groups (60% at 40 mg/day, 62% at 50 mg/day), but less drug-related serious adverse events (grade >3) occurred at the lower dose. Hence, 40 mg/day was recommended for phase III trials.

Clinical Efficacy: Comparison with Cytotoxic Chemotherapy

Afatinib significantly improved progression-free survival as a first-line treatment in patients with advanced NSCLC whose tumors harbored EGFR mutations compared to chemotherapy, according to the results of phase III LUX-Lung-3 (afatinib *versus* pemetrexed plus cisplatin) [39] and LUX-Lung-6 (afatinib *versus* gemcitabine plus cisplatin) trials [40]. The median progression free survival for the whole population was significantly longer in the afatinib group than in the chemotherapy group in both LUX-Lung 3 (11.1 *versus* 6.9 months) and LUX-Lung 6 (11.0 *versus* 5.6 months) (Table **4**). Additionally, progression-free survival improvement in the afatinib group compared to chemotherapy group for the whole population was similar as that found for subgroup of patients with common EGFR mutations (Del 19 and L858R) in both LUX-Lung 3 (13.6 *versus* 6.9 months) and LUX-Lung 6 (11.0 *versus* 5.6 months) Table (**6**).

Table 6. Clinical efficacy of afatinib as a first-line treatment for patients with EGFR-mutated NSCLC.

Trial	Treatment	No. of patients	Median PFS (months)	Median OS (months)	Median TTF (months)	Reference
Comparison with chemotherapy						
Whole population						
LUX-Lung 3	AFA	230	11.1 (HR 0.58, p<0.001)	28.2 (HR 1.12, p = 0.60)		[39]
	CT[a]	115	6.9	28.2		
LUX-Lung 6	AFA	242	11.0 (HR 0.28, p<0.0001)	23.1 (HR 0.95, p = 0.76)		[40]
	CT[b]	122	5.6	23.5		
Patients with common EGFR mutations (Del 19 or L858R)						
LUX-Lung 3	AFA	203	13.6 (HR 0.47, p<0.001)	31.6		
	CT	104	6.9	28.2		
LUX-Lung 6	AFA	216	11.0 (HR 0.25, p<0.0001)	23.6		
	CT	155	5.6	23.5		
Patients with Del 19						
Pooled analysis[c]	AFA	246		31.7 (HR 0.59, P = 0.0001)		[41]
	CT	119		20.7		
Patients with L858R						
Pooled analysis[c]	AFA	183		22.1 (HR 1.25, p = 0.16)		[41]
	CT	93		26.9		

(Table 6) cont.....

Trial	Treatment	No. of patients	Median PFS (months)	Median OS (months)	Median TTF (months)	Reference
Comparison with gefitinib						
LUX-Lung 7	AFA	160	11.0 (HR 0.73, p = 0.017)	27.9 (HR 0.87, p =0.33)	13.7	[42]
	GEF	159	10.9	24.5	11.5	

PFS: Progression-Free Survival; OS: Overall Survival; TTF: Time-to-Treatment Failure AFA: Afatinib; CT: Chemotherapy; GEF: Gefitinib [a] CT for LUX-Lung 3 includes pemetrexed + cisplatin [b] CT for LUX-Lung 6 includes gemcitabine + cisplatin [c] Combined data of LUX-Lung 3 and LUX-Lung 6

Overall survival did not significantly differ between afatinib group and chemotherapy group for whole population in either LUX-Lung 3 (28.2 and 28.2 months, respectively) or LUX-Lung 6 trials (23.1 and 23.5 months, respectively), or for the subgroup of patients with common EGFR mutations (Del 19 and L858R) in either LUX-Lung 3 (31.6 and 28.2 months, respectively) or LUX-Lung 6 trial (23.6 and 23.5 months, respectively). However, and interestingly, afatinib as a first-line treatment significantly prolonged overall survival only in patients with Del 19 but not in those with L858R compared to chemotherapy. In subgroup analysis examining different EGFR mutations, the patient group with Del 19 in afatinib demonstrated a significant improvement in overall survival compared to those in chemotherapy group in a pooled analysis of LUX-Lung 3 and LUX-Lung 6 (31.7 *versus* 20.7 months) and even in separate analysis (LUX-Lung 3: 33.3 *versus* 21.1 months, LUX-Lung 6: 31.4 *versus* 18.4 months). In contrast, patients with L858R did not show different overall survival between afatinib group and chemotherapy group in either pooled analysis (22.1 and 26.9 months, respectively) [41] or individual analysis (LUX-Lung 3: 27.6 and 40.3 months, respectively; LUX-Lung 6: 19.6 and 24.3 months, respectively) of the trials (Table **6**). However, other studies, including LUX-Lung 7 [42], have not confirmed these differences. In the LUX-Lung trial afatinib shows a benefit over gefitinb in both mutation subtypes. Thus, currently tailoring treatment with afatinib on the basis of mutation subtype (exon 19 deletion or L858R) is not indicated.

It is interesting to note that no overall survival benefit was observed in most phase III trials assessing the first-line treatment of EGFR TKIs (including gefitinib, erlotinib and afatinib) over standard platinum based chemotherapy [11, 42, 43]. Differences in overall survival can be confounded by crossover, *e.g.*, patients in the chemotherapy arms received TKIs as second- or third-line therapy after disease progression, which make it difficult to distinguish the effect from the original and subsequent therapy [44]. The high degree of crossover may extend the benefit associated with the administration of TKIs (as second- or third-line treatment) to patients originally assigned to the control or chemotherapy arm.

Clinical Efficacy: Comparison with Gefitinib

The LUX-Lung 7 trial demonstrated that afatinib was more efficacious than gefitinib as a first-line treatment in treatment-naïve patients with EGFR-mutated NSCLC [42, 45]. This open-label, phase IIb trial recruited 390 patients with common EGFR mutations (Del 19 and L858R) to randomly receive (1:1) either afatinib 40 mg/day or gefitnib 250 mg/day. The co-primary endpoints were progression-free survival, time-to-treatment failure and overall survival. Afatinib compared to gefitinib improved progression free survival (11.0 *vs* 10.9 months), time-to-treatment failure (13.7 *versus* 11.5 months), and overall survival (27.9 *versus* 24.5 months) [42, 45]. Additionally, the objective responses according to EGFR mutation types were better in afatinib group than gefitinib group (Del 19: 66% *vs* 42%, L858R: 73% *vs* 66%). However, the results of this trial were limited as it aimed at broadly examining the differences between afatinib and gefitinib without a formally predefined hypothesis and adjustment for multiple testing on the three co-primary endpoints (Table **6**).

Safety and Tolerability

Afatinib is generally well tolerated. Similar to first generation of EGFR-TKIs, the common drug-related adverse events are diarrhea and skin rash.

Quality of Life Analysis

Patient-reported symptom and health-related quality of life (HRQoL) benefit of afatinib over placebo plus best supportive care have been investigated in a randomized phase IIb/III trial (LUX-Lung 1) [46]. The analysis was conducted by the QLQ-C30/LC13 QoL analysis mentioned above for gefitinib and the EuroQoL (EQ-5D) questionnaires. The addition of afatinib to best supportive care was found to significantly improve the three most commonly studied lung cancer-related symptoms (cough, dyspnea and pain) with a dramatic delayed time to deterioration of cough.

DACOMITINIB

The optimized dose of dacomitinib is 45 mg/day, according to three phase I trials conducted in the U.S [47], South Korea [48] and Japan [49], respectively.

Clinical Efficacy as a First-line Treatment

Dacomitinib had encouraging clinical efficacy for treatment of patients with advanced EGFR-mutant NSCLC. ARCHER (Advanced Research for Cancer targeted pan-HER therapy) 1017 was a single-arm phase II study that investigated the efficacy of dacomitinib in 89 treatment-naïve patients who were never-smokers or former light smokers, or who harbored EGFR-activating mutations (Del 19 or L858R) [50]. The median progression-free survival was higher in EGFR-mutant population than in the whole population (18.2 *versus* 11.5 months). Similarly, the objective response rates were higher in EGFR-mutant population than in the whole population (75.6% *vs* 53.9%). Neither the median PFS nor objective responses differed between Del 19 and L858R mutations. Notably, however, the starting dose of dacomitinib, 45 mg /day was reduced to 30 mg/day in some patients due to intolerable toxicities, which contradicted with the recommended tolerated dose determined in phase I studies. This difference may be due to the better ability in managing EGFR-TKIs toxicities in the EGFR-TK--pretreated patients of phase I studies than in the treatment-naïve patients of the phase II trial. To verify whether dacomitinib is beneficial for advanced EGFR-mutant NSCLC patients, a phase III trial (ARCHER 1050) has been conducted to compare dacomitinib with gefitinib (Table **7**). Recently, results from the ARCHER 1050 has been published showing a benefit in progression-free survival for dacomitinib [51].

Table 7. Clinical trials of dacomitinib for advanced NSCLC.

Trial	Phase	Status	Treatment	No. of patients	ORR (%)	Median PFS (months)	Median OS (months)	Reference
First-line treatment								
ARCHER 1017								[50]
Overall population	II	Completed	DAC	89	53.9	11.5 (95% CI: 9.0-12.9)	29.5 (95% CI: 22.8-35.6)	
EGFR-mutant (Del19 or L858R)			DAC	45	75.6	18.2 (95% CI: 12.8-23.8)	40.2 (29.0 – not reached)	
ARCHER 1050								[51]
Overall population	III	Completed	DAC	227		14.7 (HR 0.59, p < 0.0001)		
			GEF	225		9.2		
Second-line treatment								
ARCHER 1028								[52]

(Table 7) cont.....

Trial	Phase	Status	Treatment	No. of patients	ORR (%)	Median PFS (months)	Median OS (months)	Reference
Overall population	II	Completed	DAC	94	17.0	2.86 (HR 0.66, p = 0.012)	9.53 (HR 0.80, p = 0.205)	
			ERL	94	5.3	1.91	7.44	
KRAS WT/EGFR WT			DAC			2.21		
			ERL			1.84		
ARCHER 1009								[53]
Overall population	III	Completed	DAC	439	11[a]	2.6[a] (HR 0.94, p=0.2879)	7.9 (HR 1.079, 95% CI: 0.914-1.274)	
			ERL	439	8[a]	2.6[a]	8.4	
KRAS WT			DAC	256	13[a]	2.6[a] (HR1.02, p=0.8551)	8.1 (HR 1.095, 95% CI: 0.882-1.360)	
			ERL	263	29[a]	2.6[a]	8.5	
KRAS WT/EGFR WT			DAC	219		1.9[a] (HR 1.01, p=0.3298)	7.0	
			ERL	221		1.9[a]	7.7	
Third-line or beyond treatment								
ARCHER1002								[54]
Overall population (patients with KRAS WT or known EGFR sensitizing mutations)	II	Completed	DAC	66	5.2	12 w[b] (95% CI:8-20)	37 w[b] (95% CI:6-18)	
EGFR-mutant			DAC	25	8	18 w[b] (95% CI: 6-30)	57 w[b] (95% CI: 2 - unreached)	
ARCHER BR-26								[55]
Overall population	III	Completed	DAC	480	7	2.66 (HR 0.66, p<0.0001)	6.83 (HR 1.00, 95% CI:0.83-1.21)	

(Table 7) cont.....

Trial	Phase	Status	Treatment	No. of patients	ORR (%)	Median PFS (months)	Median OS (months)	Reference
			Placebo	240	1	1.38	6.31	
EGFR mutated *vs* EGFR WT			DAC	114 *vs* 235		3.52 *vs* 1.91 (HR 1.34, p=0.005)	7.23 *vs* 6.93 (HR 0.98, p=0.461)	
			Placebo	68 *vs* 114		0.95 *vs* 1.63	7.52 *vs* 5.55	
KRAS mutated *vs* KRAS WT			DAC	57 *vs* 220		1.61 *vs* 3.06 (HR 0.48.1, p=0.029)	5.82 *vs* 7.00 (HR 2.10, p=0.984)	
			Placebo	21 *vs* 120		1.86 *vs* 1.05	8.28 *vs* 5.19	

ORR: Objective Response Rate; PFS: Progression-Free Survival; OS: Overall Survival; WT: Wild-Type; DAC: Dacomitinib; GEF: Gefitinib; ERL: Erlotinib [a] Confirmed by independent review [b] w: weeks

Clinical Efficacy as a Second-line Treatment

Dacomitinib was not superior to erlotinib as a second-line treatment for patients with advanced NSCLC pretreated with chemotherapy. Despite the improved progression-free survival in KRAS wild-type/EGFR wild-type in the dacomitinib group compared to erlotinib group in a phase II study (ARCHER 1028) [52], there were no differences in objective response rate, progression free survival and overall survival between dacomitinib group and erlotinib group in the whole population, or in patients with KRAS wild-type/EGFR wild-type in a phase III trial (ARCHER 1009) [53] (Table **7**).

Clinical Efficacy as a Third-line or Beyond Treatment

Dacomitinib was not recommended as a third-line or beyond treatment for patients with advanced NSCLC previously treated with chemotherapy and an EGFR-TKI. A single-arm phase II study (ARCHER 1002) of dacomitinib was conducted in 878 patients with advanced NSCLC who failed chemotherapy and erlotinib treatment and who harbor KRAS wild-type or known EGFR sensitizing mutation to investigate the role of dacomitinib in KRAS wild-type and EGFR-mutated patients [54]. Results showed that the objective response rate and progression-free survival in KRAS wild-type NSCLC patients in the dacomitinib group were similar to those in the erlotinib or gefitinib group, but the objective response rate in EGFR-mutant patients was very low. Furthermore, a phase III trial (ARCHER BR-26) assigned patients (2:1) with advanced NSCLC previously treated with both chemotherapy and a first-generation EGFR-TKI to receive either dacomitinib or placebo [55]. Disappointingly, dacomitinib did not improve overall survival (6.83 and 6.31 months for dacomitinib and placebo group, respectively). Moreover, there were no significant difference in overall survival according to KRAS and EGFR status (Table **7**).

Quality of Life Analysis

The health-related quality of life (HRQoL) of dacomitinib in NSCLC patients with KRAS wild-type tumor who progressed after chemotherapy regimens and erlotinib was evaluated in a phase II trial [54]. The 30-question EORTC QLQ-C30 questionnaire core module and the 13-item lung cancer symptom-specific module QLQ-LC13 were used in the assessment. The impact of dacomitinib on patients' skin condition was also evaluated using a 10-item Dermatology Life Quality Index (DLQI) questionnaire. The QoL scores reported by the patients were at their worst in the mid-range. NSCLC symptoms of dyspnea, cough, and pain showed improvement after 3 weeks on therapy.

CANERTINIB

Canertinib is currently not developed for the treatment of advanced NSCLC due to its disappointing clinical outcomes [56]. A phase II trial evaluated the efficacy of canertinib at three dose levels (50 mg/day, 150 mg/day and 450 mg/day) in patients with advanced stage NSCLC who failed platinum-based chemotherapy. However, there were no significant differences in 1-year survival rate, progression free survival and overall survival among the three groups. In addition, there were no complete responses and only few partial responses (2%, 2% and 4% at the three dose levels, respectively). Hence, clinical development of canertinib for treating NSCLC is not further pursued.

It was speculated that the unsatisfactory clinical response may be partially contributed to by the patient composition in the trial. There was a relatively higher percentage of patients with squamous cell carcinoma in the canertinib trial than in other gefitinib and erlotinib studies. Squamous cell carcinoma is known to produce poor radiographic response, which was the assessment method for efficacy in canertinib trial. The higher frequency of patients with squamous cell carcinoma in the canertinib trial also suggests that the patients are likely former or current smokers, who generally provide a poorer response to EGFR TKI [57]. Moreover, no patient of Asian ethnicity was recruited in the canertinib trial. However, Asian patients are known to respond better to EGFR TKIs.

NERATINIB

Neratinib is no longer under clinical investigation for the treatment of NSCLC because of its insufficient bioavailability from diarrhea-imposed dose limitation. Its maximum tolerated dose is 320 mg/day, according to a phase I study conducted in patients with solid tumors [58]. However, during phase II trial, the dose of neratinib was reduced to 240 mg/day due to excessive diarrhea (grade 3 > 50%) [59]. Unfortunately, 240 mg/day was too low for patients to achieve the therapeutic level required to effectively inhibit EGFR Del 19 and T790M mutations *in vivo* [60].

DRUG RESISTANCE TO SECOND GENERATION EGFR TKI (AFATINIB)

There have been only a few clinical investigations into the mechanisms of acquired resistance to afatinib [61 - 63]. Interestingly, the general consensus is that EGFR T790M mutation is indeed the major mechanism of resistance to afatinib, akin to that causing drug resistance to the first-generation EGFR TKIs [63]. Therefore, afatinib-treated patients are equally good candidates as those treated with 1st generation EGFR TKIs when considering novel 3rd generation T790M-specific inhibitors (see later) [64]. In preclinical models, afatinib inhibits EGFR T790M mutant protein [65], but the clinical use of afatinib as salvage therapy after first generation EGFR TKIs has been disappointing [66, 67]. It was found that the irreversible 2nd generation EGFR TKIs are not effective against the EGFR T790M resistant cells *in vivo* at a clinically achievable concentration [68, 69].

Recently, besides the T790M secondary mutation, two additional EGFR mutants (L792F and C797S) were identified from *in vitro* study that lead to afatinib resistance [70]. The amplification or overexpression of wild-type KRAS and increased expression of insulin-like growth factor binding protein 3 (IGFBP3) were also reported in *in vitro*-induced afatinib resistant lung adenocarcinoma cell lines, indicating a potential bypass signaling pathway [71]. However, the clinical relevance of these novel mutations and mechanisms has yet to be established in clinical investigation.

THIRD-GENERATION OF EGFR-TKIS

Third-generation EGFR-TKIs include osimertinib (AZD9291, mereletinib, Tagrisso™), rociletinib (CO-1686, AVL-301), olmutinib (Olita™, HM61713, BI 1482694), ASP8273, EGF816 and PF-06747775 (Table **1**). Thus far, only osimertinib has received FDA approval on 10 November 2015 and EMA approval on 25 April 2017 for the treatment of advanced EGFR T790M-positive NSCLC. Olmutinib was approved in South Korea on 17 May 2016 for the treatment of patients with locally advanced or metastatic EGFR T790M mutation-positive

NSCLC, who had been previously treated with an EGFR-TKI. The clinical development of rociletinib has been stopped after a series of negative developments for the drug, including updated data revealing lower response rates than initially reported.

STRUCTURES AND MECHANISM OF ACTION

The structures of osimertinib, rociletinib, olmutinib, all of which are low molecular weight anilinopyrimidine derivatives, are showed in Fig. (**4**). Third-generation EGFR-TKIs act as irreversible inhibitors that target T790M mutation while sparing wild-type EGFR. They bind covalently (Cys 797 of EGFR) to the certain forms of EGFR (Del 19, L858R and double mutants containing T790M mutation), and thus inhibiting the downstream signaling pathway.

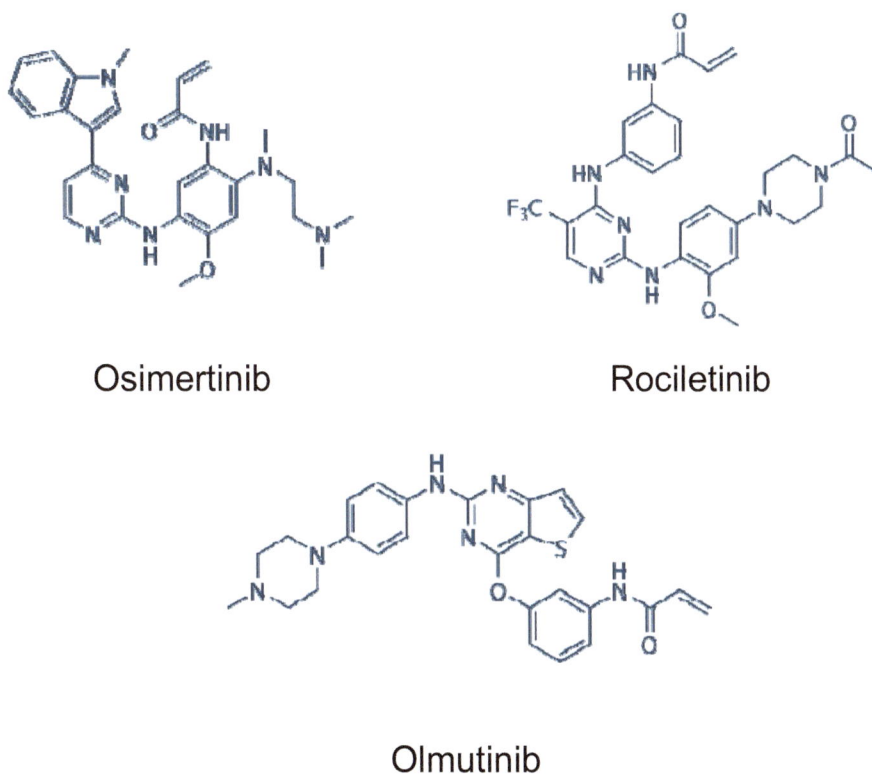

Osimertinib   Rociletinib

Olmutinib

Fig. (4). Chemical structures of third-generation EGFR TKIs.

OSIMERTINIB

The optimized dose of osimertinib is 80 mg/day. A phase I study (AURA) was conducted to evaluate the safety and efficacy of osimertinib in patients with advanced EGFR-mutated NSCLC which progressed on a first- or second-generation EGFR-TKI. In the dose-escalation cohort, the patients received 20, 40, 80, 160 or 240 mg/day of osimertinib. Since the tumor response rates were similar among the five dosages with increasing toxicities at 160 and 240 mg/day, osimertinib 80 mg/day was recommended for further clinical development [64].

Clinical Efficacy

Osimertinib showed encouraging efficacy in patients with EGFR-TKI sensitizing and T790M mutation positive advanced NSCLC. In AURA phase I trial, a subgroup of patients with advanced NSCLC who harbored EGFR-TKI sensitizing and T790M mutations treated with 80 mg/day osimertinib provided a high objective response rate (71%), with encouraging median duration of response (9.6 months) and progression free survival (9.7 months) [72]. These findings were confirmed by a pooled analysis of two phase II studies (AURA phase II extension cohort and AURA 2), where 411 patients with advanced NSCLC patients harboring EGFR-TKI sensitizing and T790M mutation received 80 mg/day of osimertinib. Results showed that objective response rate was 66%, median duration of response was 12.5 months and median progression free survival was 11.0 months [72] (Table **8**).

Table 8. Key clinical trials of osimertinib monotherapy.

	AURA phase I	**Pooled phase II** **(AURA phase II extension + AURA 2)** **[72]**
No. of patients	63	411
Objective response rate (%)	71	66
Median duration of response (months)	9.6	12.5
PFS (months)	9.7 (95% CI: 8.3-13.6)	11.0 (95% CI: 9.6-12.4)

Numerous phase I, II and III trials are ongoing to further evaluate the safety and efficacy of osimertinib as a first-line or second-line treatment for advanced EGFR T790M-mutated NSCLC compared to chemotherapy and other EGFR-TKIs (Table **9**). In the AURA3 study (NCT02151981; a randomized, open-label, phase III trial in patients with T790M-positive advanced NSCLC) [73], osimertinib was showed to significantly prolong progression-free survival and give rise to better objective response rate than platinum therapy plus pemetrexed. In a more recent FLAURA study (NCT02296125; a double-blind, phase III trial in previously untreated patients with EGFR mutation-positive (exon 19 deletion or L858R) advanced NSCLC) [74], osimertinib gave rise to statistically significant longer progression-free survival than standard EGFR TKIs (18.9 *versus* 10.2 months) in the first-line treatment setting, with a similar safety profile and lower rates of serious adverse events [74].

Table 9. Ongoing clinical trials of osimertinib monotherapy.

Phase	Title	Intervention/contents	NCT no.
Treatment in first-line setting			
II	Osimertinib (AZD9291) in First-line Locally Advanced or Metastatic NSCLC Patients With EGFR and EGFR T790M	Single-armed	02841579
II	Osimertinib Treatment on EGFR T790M Plasma Positive NSCLC Patients (APPLE)	Osimertinib versus gefitinib	02856893
II	Neo-adjuvant Trial With AZD9291 in EGFRm+ Stage IIIA/B NSCLC	Single-armed	
III	AZD9291 Versus Placebo in Patients With Stage IB-IIIA Non-small Cell Lung Carcinoma, Following Complete Tumour Resection With or Without Adjuvant Chemotherapy.	Osimertinib versus placebo	02511106
III	AZD9291 Versus Gefitinib or Erlotinib in Patients With Locally Advanced or Metastatic NSCLC	Osimertinib versus Gefitinib / Erlotinib	02296125
Treatment in second or third-line setting			
II	Phase II Study of AZD9291 in Advanced Stage NSCLC With EGFR and T790M Mutations Detected in Plasma Ct-DNA	Single-armed	02811354
II	Phase II AZD9291 Open Label Study in NSCLC After Previous EGFR TKI Therapy in EGFR and T790M Mutation Positive Tumors	Single-armed	02094261
II	Phase II Single Arm Study of AZD9291 to Treat NSCLC Patients in Asia Pacific	Single-armed	02442349
II	AZD9291, an Irreversible EGFR-TKI, in Relapsed EGFR-mutated NSCLC Patients Previously Treated With an EGFR-TKI, Coupled to Extensive Translational Studies	Single-armed	02504346
II/III	Osimertinib or Docetaxel-bevacizumab as Third-line Treatment in EGFR T790M Mutated NSCLC	Osimertinib versus Docetaxel-bevacizumab	02959749
III	AZD9291 Versus Platinum-Based Doublet-Chemotherapy in Locally Advanced or Metastatic NSCLC	Osimertinib versus pemetrexed+carboplatin/ cisplatin	02151981

(Table 9) cont.....

Phase	Title	Intervention/contents	NCT no.
III	Real World Treatment Study of AZD9291 for Advanced/Metastatic EGFR T790M Mutation NSCLC	Single-armed	02474355
Observation trial			
--	KOREA Study (Korea Osimertinib Real World Evidence Study to Assess Safety and Efficacy)	Proportion and severity of adverse events (primary outcome measures)	02777567
--	Observational Study of Patients With Locally Advanced or Metastatic NSCLC	Several molecular testing parameters and clinical outcomes	03053297
--	Tagrisso Tablets Clinical Experience Investigation (All Case Investigation)	Incidence of adverse drug reactions, factors which may affect safety and efficacy and information of ADRs not expected from "Precautions for Use" of the package insert in Japan	02756039

Clinical Use in Patients with Leptomeningeal Metastases (LM)

Metastasis of NSCLC to the brain is common and it is associated with poor prognosis. In the BLOOM study [75], the clinical efficacy was evaluated in NSCLC patients with advanced mutated EGFR disease who had progressed after prior EGFR TKI treatment and confirmed with LM. The penetration of osimertinib into the brain was confirmed. Promising clinical response and manageable adverse effect were observed with a median treatment duration of 6 months. The clinical trial is still on going and final data has not been posted yet. Other studies showing the CNS-activity of osimertinib includes a pooled analysis from the two phase II, AURA extension (NCT01802632) and AURA2 (NCT02094261) [76]. Osimertinib was showed to exhibit significant clinical efficacy against CNS metastases, with an encouraging objective response rate and a high disease control rate were 54% (27/50; 95% CI 39-68) and 92% (46/50; 95% CI 81-98), respectively. The efficacy of osimertinib in CNS metastases was also demonstrated in a randomized phase III study (AURA3; NCT02151981) in patients with T790M-positive advanced NSCLC who have progressed on or after prior EGFR TKI therapy [73]. In the CNS evaluable for response set of patients (which included only patients with >1 measurable CNS metastases), CNS objective response rate was 70% (95% CI 51-85) with osimertinib therapy and only 31% (95% CI 11-59) with chemotherapy (OR 5.13; 95% CI 1.44-20.64; p = 0.015) [73]. Median CNS progression-free survival was also significantly longer with osimertinib therapy than with chemotherapy (11.7 *versus* 5.6 months; HR 0.32; 95% CI 0.15-0.69; p = 0.004) [73].

Safety and Tolerability

Osimertinib was well-tolerated in patients with advanced EGFR-mutated NSCLC. The MTD was not defined, as no dose-limiting toxicities were observed at dose level ranging from 20 mg/day to 240 mg/day. In the pooled analysis of two phase II studies, the most common drug-related adverse events are diarrhea (41%), rash (38%). Only few patients (<1%) had grade 3 or higher drug-related adverse events [72].

Quality of Life (QoL) Analysis

Health-related QoL is being assessed as a secondary endpoint in a phase III trial (AURA3), which compared the efficacy and safety of osimertinib and platinum-based doublet chemotherapy in NSCLC patients with EGFR T790M mutation, locally advanced or metastatic disease whose disease had progressed on or after treatment with a previous EGFR TKI. The trial is carried out in more than 130 sites worldwide. The data from AURA3 study has been reported [73]. From the repeated-measures analysis, patient-reported outcomes were better in the osimertinib group than in the platinum-pemetrexed group in all five pre-specified symptoms. Also, improvement of lung cancer-related symptoms (including appetite loss, fatigue, breathlessness and chest pain) with osimertinib was reported at the European Lung Cancer Conference 2017 [77].

ROCILETINIB

Rociletinib was found to be effective in patients with advanced EGFR T790M mutated NSCLC as a second-line treatment (Table **10**). A phase I/II study was conducted in patients with advanced EGFR-mutated NSCLC who progressed after a first- or second-generation EGFR-TKI treatment [78]. In phase I, patients received either rociletinib of free-base (started from 150 mg/day to 900 mg twice daily) or HBr form (started from 500 mg to 1000 mg twice daily). MTD was not defined. The most common dose-limiting adverse event was hyperglycemia. In phase II, in patients who are T790M positive compared to those who are T790M-negative, rociletinib showed significant higher objective response rate (59% *versus* 29%), disease control rate (defined as the proportion of patients with a complete or partial response or stable disease, 93% *versus* 59%) and progression free survival (13.1 *versus* 5.6 months). A confirmatory phase II trial of rociletinib (625 mg twice a day) as a second-line treatment for advanced EGFR-mutated NSCLC who progressed after an EGFR-TKI treatment (TIGER-2) was also conducted. Additionally, several clinical trials have been performed to assess the efficacy and safety of rociletinib in single-armed studies (TIGER-2 and NCT01526928) and in comparison with chemotherapy (TIGER-3) and with erlotinib (TIGER-1). Rociletinib (more precisely, its metabolite M502) was also reported in preclinical study to inhibit insulin-like growth factor 1 receptor (IGF1R) and insulin receptor (INSR) [79], the activation of which is known to contribute to resistance to EGFR TKIs [80]. However, all ongoing clinical studies on rociletinib were halted in May 2016 because new drug application for rociletinib was not approved by the FDA.

OLMUTINIB

Olmutinib showed promising clinical activity in an ongoing phase I/II study (NCT01588145) clinical trial. At the recommended dose of 800 mg/day, 71 Korean patients with EGFR T790M mutation-positive advanced NSCLC (who developed resistance to initial EGFR TKI therapy) exhibited an objective response rate of 56% with a median duration of response of 8.3 months. Disease control rate was 90% and median progression free survival was 7.0 months. Majority of treatment-related adverse events were mild-to-moderate and the most common included diarrhea, nausea, skin rash and skin itching. Olmutinib is being further evaluated in a global phase II ELUXA 1 trial and other phase I and II studies (Table **11**).

ASP8273

ASP8273 is currently under clinical investigation. A phase I/II study was conducted in Japanese patients to evaluate the safety and efficacy of ASP8273 [81]. Maximum tolerated dose was determined to be 400 mg. Partial response rate was 50% in whole population and 80% in patients with T790M. Results showed that 300 mg/day was the recommended phase II dose. Several phase I, II and III clinical trials are ongoing to further elucidate its safety, pharmacokinetics, and clinical efficacy (Table **12**). On 10 May 2017, Astellas announced the discontinuation of ASP8273 treatment arm in the the late-stage SOLAR trial (NCT02588261) evaluating the efficacy and safety of ASP8273 *versus* erlotinib/gefitinib for the 1st line treatment metastatic or advanced unresectable non-small cell lung cancer (NSCLC). Furthermore, no new patients are being enrolled in ASP8273 trials.

Table 10. Clinical trials investigating rociletinib monotherapy before the termination of its development.

Phase	Title	Intervention/contents	NCT no.
	Treatment in first-line setting		
II/III	**TIGER-1:** Safety and Efficacy Study of Rociletinib (CO-1686) or Erlotinib in Patients With EGFR-mutant/Metastatic NSCLC Who Have Not Had Any Previous EGFR Directed Therapy	Rociletinib versus erlotinib	02186301

(Table 10) cont.....

Phase	Title	Intervention/contents	NCT no.
	Treatment in second or third-line setting		
I/II	Study to Evaluate Safety, Pharmacokinetics, and Efficacy of Rociletinib (CO-1686) in Previously Treated Mutant EGFR in NSCLC Patients	Phase II: single-armed	01526928
II	**TIGER-2:** A Phase 2, Open-label, Multicenter, Safety and Efficacy Study of Oral CO-1686 as 2nd Line EGFR-directed TKI in Patients With Mutant EGFR NSCLC	Single-armed	02147990
III	**TIGER-3:** Open Label, Multicenter Study of Rociletinib (CO-1686) Mono Therapy Versus Single-agent Cytotoxic Chemotherapy in Patients With Mutant EGFR NSCLC Who Have Failed at Least One Previous EGFR-Directed TKI and Platinum-doublet Chemotherapy	Rociletinib versus Pemetrexed/gemcitabine/ paclitaxel/docetaxel	02322281

Table 11. Clinical trials of olmutinib monotherapy before the termination of its development because of serious side effect.

Phase	Title	Intervention/contents	NCT no.
	Safety, tolerability, pharmacodynamics and pharmacokinetics study		
I	Study to Evaluate a Pharmacokinetic of HM61713 in Healthy Male Subjects	Pharmacokinetics	01894399
	Treatment in first-line setting		
II	Phase II Trial to Evaluate the Efficacy and Safety of HM61713 as the 1st-line NSCLC Anticancer Therapy	Single-armed	02444819
	Treatment in second or third-line setting		
I/II	Phase I/II Trial to Evaluate Safety, Tolerability and Pharmacokinetic Profile of HM61713 in NSCLC Patients	Phase II: single-armed	01588145
II	**ELUXA 1:** Phase II Trial of HM61713 for the Treatment of ≥2nd Line T790M Mutation Positive Adenocarcinoma of the Lung	Single-armed	02485652

Table 12. Ongoing clinical trials of ASP8273 monotherapy.

Phase	Title	Intervention/contents	NCT no.
	Safety, tolerability, pharmacodynamics and pharmacokinetics study		
I	A Study of ASP8273 in Subjects With NSCLC Harboring EGFR Mutations	Bioequivalence, safety and tolerability	03082300
I	A Study to Investigate the Absorption, Metabolism and Excretion of [14C] ASP8273 in Subjects With Solid Tumors	Pharmacokinetics	02674555

(Table 12) cont.....

Phase	Title	Intervention/contents	NCT no.
I	A Dose Escalation Study of ASP8273 in Subjects With NSCLC Who Have EGFR Mutations	Dose escalation	02113813
Treatment in first-line setting			
III	A Study of ASP8273 *vs.* Erlotinib or Gefitinib in First-line Treatment of Patients With Stage IIIB/IV NSCLC Tumors With EGFR Activating Mutations	ASP8273 versus gefitinib/erlotinib	02588261
II	A Study of ASP8273 in EGFR-TKI-Naïve Patients With NSCLC Harboring EGFR Mutations	single-armed	02500927
Treatment in second or third-line setting			
I/II	An Open Study of ASP8273 in Patients With NSCLC Who Have EGFR Mutations	Phase II: single-armed	02192697
II	A Study for Subjects Who Are Participating in an Astellas-sponsored ASP8273 Study	single-armed	03042013

EGF816

A phase I multicenter, dose escalation trial was conduct to evaluate the safety, tolerability and antitumor activity of EGF816 in patients with advanced EGFR T790M mutated NSCLC. The patients were treated across 6 cohorts from 75 to 350 mg for capsules and 225 mg for tablets. The most common adverse events were diarrhea, stomatitis rash and pruritus. The objective response rate and disease control rate were 54.5% and 86.4%, respectively. A Phase I/II, multicenter, open-label study of EGF816 is ongoing to assess the safety and efficacy in patients with EGFR-mutated solid malignancies (NCT02108964).

PF-06747775

A phase I/II open-label study (NCT02349633) is ongoing to assess the safety, pharmacokinetic, pharmacodynamics and anti-tumor activity of PF-06747775 as a single agent in patients with advanced NSCLC with EGFR mutations (Del19/L858R +/- T790M). The starting dose for dose escalation study is 25 mg/day.

DRUG RESISTANCE TO THIRD-GENERATION EGFR TKI (OSIMERTINIB)

Although encouraging survival data and response rates were observed in patients treated with the third-generation EGFR TKI (osimertinib), unfortunately acquired resistance has already been reported after about 10 months of treatment. Mechanisms leading to disease progression are various and are not fully understood. EGFR-dependent mechanisms include new tertiary mutations (C797S and others) [82], EGFR gene amplification [83], and reduction/disappearance of T790M cell clones [83]. Alternative signaling pathways may be activated to cause osimertinib resistance as evidenced in clinical and/or preclinical studies. These include HER2 and MET amplification, PIK3CA activating mutations, PTEN deletion and RAS mutations [84]. Moreover, histological transformation (small cell lung cancer (SCLC) transformation and epithelial-mesenchymal transition (EMT)) has also been reported to cause resistance to third-generation EGFR TKIs [85 - 87].

CLINICAL MODES OF DISEASE PROGRESSION IN EGFR TKI-TREATED PATIENTS AND SUBSEQUENT MANAGEMENT

The National Comprehensive Cancer Network (NCCN) guideline for NSCLC version 5.2017 has developed treatment strategies for lung cancer patients with different patterns of disease progression after EGFR TKI failure according to T790M status and patients' symptomatic burden. Asymptomatic patients (*i.e.*, gradual progression) may continue previous TKI or turn to osimertinib (if T790M is confirmed). For patients with local progression, the recommendation is to continue previous TKI or switch to osimertinib (if T790M positive) and supplement with local interventions such as radiotherapy. Patients suffering from severe symptoms after acquired resistance to prior EGFR TKI (*i.e.*, dramatic progression) are recommended to switch to chemotherapy or osimertinib (if T790M positive). If brain metastasis is involved, NCCN guidelines for CNS cancers should be followed. Since the first-generation EGFR TKIs (gefitinib and erlotinib) have low penetration across the blood brain barrier [88, 89], they are not recommended for treatment of dramatically progressing patients with brain metastasis. Afatinib and osimertinib (if T790M positive) are reasonable choices.

CONCLUDING REMARKS

The discovery of EGFR mutations and the success story of EGFR TKI therapy have changed the paradigm of cancer chemotherapy from empirical cytotoxic regimens to molecular targeted therapy. It has also made possible personalized medicine in cancer treatment. EGFR TKI therapy has now become the standard treatment for EGFR mutant patients as first-line therapy. Unfortunately, EGFR-dependent and –independent mechanisms have emerged to mediate drug resistance to EGFR TKIs, which highlight the importance of repeat tumor biopsies and/or to collect plasma circulating tumor DNA at the time of disease progression. To this end, liquid biopsy test offers a convenient and minimally invasive means to determine the suitability of patients for EGFR-targeted therapy by analysis of circulating tumor DNA in peripheral blood samples. It is also useful in monitoring disease progression. The presence of EGFR T790M mutation has been reported by the analysis of circulating tumor DNA [90]. With the availability of the EGFR T790M-seletive EGFR TKI (osimertinib) in the clinic, detection of this mutation is clinically meaningful. In fact, the cobas EGFR Mutation Test v2 has recently been approved by FDA in June 2016 as a companion diagnostic test for the detection of exon 19 deletion or exon 21 substitution mutations in the EGFR gene from plasma samples [91]. If the result is negative, the routine FEPE tissue staining of tumor biopsy is recommended. A better understanding of the mechanisms of resistance is key in the future development of next generation EGFR TKIs.

CONSENT FOR PUBLICATION

Not applicable.

CONFLICT OF INTEREST

The authors declare no conflict of interest, financial or otherwise

ACKNOWLEDGEMENT

Declared none.

REFERENCES

[1]   American Cancer Society. Cancer Facts & Figures 2016. https://www.cancer.org/research/cancer-facts-statistics/all-cancer-facts-figures/cancer-facts-figures-2016.html Accessed March 1, 2017

[2]   Travis WD, Brambilla E, Nicholson AG, *et al.* The 2015 World Health Organization Classification of Lung Tumors: Impact of Genetic, Clinical and Radiologic Advances Since the 2004 Classification. J Thorac Oncol 2015; 10(9): 1243-60.
[http://dx.doi.org/10.1097/JTO.0000000000000630] [PMID: 26291008]

[3]   Herbst RS. Review of epidermal growth factor receptor biology. Int J Radiat Oncol Biol Phys 2004; 59(2) (Suppl.): 21-6.
[http://dx.doi.org/10.1016/j.ijrobp.2003.11.041] [PMID: 15142631]

[4]   Lynch TJ, Bell DW, Sordella R, *et al.* Activating mutations in the epidermal growth factor receptor underlying responsiveness of non-small-cell lung cancer to gefitinib. N Engl J Med 2004; 350(21): 2129-39.
[http://dx.doi.org/10.1056/NEJMoa040938] [PMID: 15118073]

[5]   Mok TS, Wu YL, Thongprasert S, *et al.* Gefitinib or carboplatin-paclitaxel in pulmonary adenocarcinoma. N Engl J Med 2009; 361(10): 947-57.
[http://dx.doi.org/10.1056/NEJMoa0810699] [PMID: 19692680]

[6]   Maemondo M, Inoue A, Kobayashi K, *et al.* Gefitinib or chemotherapy for non-small-cell lung cancer with mutated EGFR. N Engl J Med 2010; 362(25): 2380-8.
[http://dx.doi.org/10.1056/NEJMoa0909530] [PMID: 20573926]

[7]   Kosaka T, Yatabe Y, Endoh H, Kuwano H, Takahashi T, Mitsudomi T. Mutations of the epidermal growth factor receptor gene in lung cancer: biological and clinical implications. Cancer Res 2004; 64(24): 8919-23.
[http://dx.doi.org/10.1158/0008-5472.CAN-04-2818] [PMID: 15604253]

[8] Paez JG, Jänne PA, Lee JC, *et al.* EGFR mutations in lung cancer: correlation with clinical response to gefitinib therapy. Science 2004; 304(5676): 1497-500.
[http://dx.doi.org/10.1126/science.1099314] [PMID: 15118125]

[9] Fukuoka M, Yano S, Giaccone G, *et al.* Multi-institutional randomized phase II trial of gefitinib for previously treated patients with advanced non-small-cell lung cancer (The IDEAL 1 Trial) [corrected]. J Clin Oncol 2003; 21(12): 2237-46.
[http://dx.doi.org/10.1200/JCO.2003.10.038] [PMID: 12748244]

[10] Mok TS, Wu YL, Thongprasert S, *et al.* Gefitinib or carboplatin-paclitaxel in pulmonary adenocarcinoma. N Engl J Med 2009; 361(10): 947-57.
[http://dx.doi.org/10.1056/NEJMoa0810699] [PMID: 19692680]

[11] Mitsudomi T, Morita S, Yatabe Y, *et al.* Gefitinib versus cisplatin plus docetaxel in patients with non-small-cell lung cancer harbouring mutations of the epidermal growth factor receptor (WJTOG3405): an open label, randomised phase 3 trial. Lancet Oncol 2010; 11(2): 121-8.
[http://dx.doi.org/10.1016/S1470-2045(09)70364-X] [PMID: 20022809]

[12] Herbst RS, Maddox AM, Rothenberg ML, *et al.* Selective oral epidermal growth factor receptor tyrosine kinase inhibitor ZD1839 is generally well-tolerated and has activity in non-small-cell lung cancer and other solid tumors: results of a phase I trial. J Clin Oncol 2002; 20(18): 3815-25.
[http://dx.doi.org/10.1200/JCO.2002.03.038] [PMID: 12228201]

[13] Oizumi S, Kobayashi K, Inoue A, *et al.* Quality of life with gefitinib in patients with EGFR-mutated non-small cell lung cancer: quality of life analysis of North East Japan Study Group 002 Trial. Oncologist 2012; 17(6): 863-70.
[http://dx.doi.org/10.1634/theoncologist.2011-0426] [PMID: 22581822]

[14] Hidalgo M, Siu LL, Nemunaitis J, *et al.* Phase I and pharmacologic study of OSI-774, an epidermal growth factor receptor tyrosine kinase inhibitor, in patients with advanced solid malignancies. J Clin Oncol 2001; 19(13): 3267-79.
[http://dx.doi.org/10.1200/JCO.2001.19.13.3267] [PMID: 11432895]

[15] Rosell R, Carcereny E, Gervais R, *et al.* Erlotinib *versus* standard chemotherapy as first-line treatment for European patients with advanced EGFR mutation-positive non-small-cell lung cancer (EURTAC): a multicentre, open-label, randomised phase 3 trial. Lancet Oncol 2012; 13(3): 239-46.
[http://dx.doi.org/10.1016/S1470-2045(11)70393-X] [PMID: 22285168]

[16] Zhou C, Wu YL, Chen G, *et al.* Erlotinib versus chemotherapy as first-line treatment for patients with advanced EGFR mutation-positive non-small-cell lung cancer (OPTIMAL, CTONG-0802): a multicentre, open-label, randomised, phase 3 study. Lancet Oncol 2011; 12(8): 735-42.
[http://dx.doi.org/10.1016/S1470-2045(11)70184-X] [PMID: 21783417]

[17] Cappuzzo F, Ciuleanu T, Stelmakh L, *et al.* Erlotinib as maintenance treatment in advanced non-small-cell lung cancer: a multicentre, randomised, placebo-controlled phase 3 study. Lancet Oncol 2010; 11(6): 521-9.
[http://dx.doi.org/10.1016/S1470-2045(10)70112-1] [PMID: 20493771]

[18] Cicènas S, Geater SL, Petrov P, *et al.* Maintenance erlotinib *versus* erlotinib at disease progression in patients with advanced non-small-cell lung cancer who have not progressed following platinum-based chemotherapy (IUNO study). Lung Cancer 2016; 102: 30-7.
[http://dx.doi.org/10.1016/j.lungcan.2016.10.007] [PMID: 27987585]

[19] Brugger W, Triller N, Blasinska-Morawiec M, *et al.* Prospective molecular marker analyses of EGFR and KRAS from a randomized, placebo-controlled study of erlotinib maintenance therapy in advanced non-small-cell lung cancer. J Clin Oncol 2011; 29(31): 4113-20.
[http://dx.doi.org/10.1200/JCO.2010.31.8162] [PMID: 21969500]

[20] Bezjak A, Tu D, Seymour L, *et al.* Symptom improvement in lung cancer patients treated with erlotinib: quality of life analysis of the National Cancer Institute of Canada Clinical Trials Group Study BR.21. J Clin Oncol 2006; 24(24): 3831-7.

[http://dx.doi.org/10.1200/JCO.2006.05.8073] [PMID: 16921034]

[21]   Chen G, Feng J, Zhou C, *et al.* Quality of life (QoL) analyses from OPTIMAL (CTONG-0802), a phase III, randomised, open-label study of first-line erlotinib versus chemotherapy in patients with advanced EGFR mutation-positive non-small-cell lung cancer (NSCLC). Ann Oncol 2013; 24(6): 1615-22.
[http://dx.doi.org/10.1093/annonc/mdt012] [PMID: 23456778]

[22]   Wang HP, Zhang L, Wang YX, *et al.* Phase I trial of icotinib, a novel epidermal growth factor receptor tyrosine kinase inhibitor, in Chinese patients with non-small cell lung cancer. Chin Med J (Engl) 2011; 124(13): 1933.
[PMID: 22088449]

[23]   Zhao Q, Shentu J, Xu N, *et al.* Phase I study of icotinib hydrochloride (BPI-2009H), an oral EGFR tyrosine kinase inhibitor, in patients with advanced NSCLC and other solid tumors. Lung Cancer 2011; 73(2): 195-202.
[http://dx.doi.org/10.1016/j.lungcan.2010.11.007] [PMID: 21144613]

[24]   Shi Y, Zhang L, Liu X, *et al.* Icotinib *versus* gefitinib in previously treated advanced non-small-cell lung cancer (ICOGEN): a randomised, double-blind phase 3 non-inferiority trial. Lancet Oncol 2013; 14(10): 953-61.
[http://dx.doi.org/10.1016/S1470-2045(13)70355-3] [PMID: 23948351]

[25]   Shi YK, Wang L, Han BH, *et al.* First-line icotinib versus cisplatin/pemetrexed plus pemetrexed maintenance therapy for patients with advanced EGFR mutation-positive lung adenocarcinoma (CONVINCE): a phase 3, open-label, randomized study. Ann Oncol 2017; 28(10): 2443-50.
[http://dx.doi.org/10.1093/annonc/mdx359] [PMID: 28945850]

[26]   Hu X, Zhang L, Shi Y, *et al.* The efficacy and safety of icotinib in patients with advanced non-small cell lung cancer previously treated with chemotherapy: a single-arm, multi-center, prospective study. PLoS One 2015; 10(11): e0142500.
[http://dx.doi.org/10.1371/journal.pone.0142500] [PMID: 26599904]

[27]   Jackman DM, Yeap BY, Sequist LV, *et al.* Exon 19 deletion mutations of epidermal growth factor receptor are associated with prolonged survival in non-small cell lung cancer patients treated with gefitinib or erlotinib. Clin Cancer Res 2006; 12(13): 3908-14.
[http://dx.doi.org/10.1158/1078-0432.CCR-06-0462] [PMID: 16818686]

[28]   Riely GJ, Pao W, Pham D, *et al.* Clinical course of patients with non-small cell lung cancer and epidermal growth factor receptor exon 19 and exon 21 mutations treated with gefitinib or erlotinib. Clin Cancer Res 2006; 12(3 Pt 1): 839-44.
[http://dx.doi.org/10.1158/1078-0432.CCR-05-1846] [PMID: 16467097]

[29]   Kumar A, Petri ET, Halmos B, Boggon TJ. Structure and clinical relevance of the epidermal growth factor receptor in human cancer. J Clin Oncol 2008; 26(10): 1742-51.
[http://dx.doi.org/10.1200/JCO.2007.12.1178] [PMID: 18375904]

[30]   Carey KD, Garton AJ, Romero MS, *et al.* Kinetic analysis of epidermal growth factor receptor somatic mutant proteins shows increased sensitivity to the epidermal growth factor receptor tyrosine kinase inhibitor, erlotinib. Cancer Res 2006; 66(16): 8163-71.
[http://dx.doi.org/10.1158/0008-5472.CAN-06-0453] [PMID: 16912195]

[31]   Oxnard GR, Arcila ME, Sima CS, *et al.* Acquired resistance to EGFR tyrosine kinase inhibitors in EGFR-mutant lung cancer: distinct natural history of patients with tumors harboring the T790M mutation. Clin Cancer Res 2011; 17(6): 1616-22.
[http://dx.doi.org/10.1158/1078-0432.CCR-10-2692] [PMID: 21135146]

[32]   Eck MJ, Yun CH. Structural and mechanistic underpinnings of the differential drug sensitivity of EGFR mutations in non-small cell lung cancer. Biochim Biophys Acta 2010; 1804(3): 559-66.
[http://dx.doi.org/10.1016/j.bbapap.2009.12.010] [PMID: 20026433]

[33]   Costa DB, Yasuda H, Sng NY. Sensitivity to EGFR inhibitors based on location of EGFR exon 20

insertion mutations within the tyrosine kinase domain of EGFR. J Clin Oncol 2012; 30: 110-2.
[http://dx.doi.org/10.1200/JCO.2011.39.4486] [PMID: 22067391]

[34]   Ji H, Zhao X, Yuza Y, *et al.* Epidermal growth factor receptor variant III mutations in lung tumorigenesis and sensitivity to tyrosine kinase inhibitors. Proc Natl Acad Sci USA 2006; 103(20): 7817-22.
[http://dx.doi.org/10.1073/pnas.0510284103] [PMID: 16672372]

[35]   Morgillo F, Della Corte CM, Fasano M, Ciardiello F. Mechanisms of resistance to EGFR-targeted drugs: lung cancer. ESMO Open 2016; 1(3): e000060.
[http://dx.doi.org/10.1136/esmoopen-2016-000060] [PMID: 27843613]

[36]   Faber AC, Corcoran RB, Ebi H, *et al.* BIM expression in treatment-naive cancers predicts responsiveness to kinase inhibitors. Cancer Discov 2011; 1(4): 352-65.
[http://dx.doi.org/10.1158/2159-8290.CD-11-0106] [PMID: 22145099]

[37]   Tong CWS, Wu WKK, Loong HHF, Cho WCS, To KKW. Drug combination approach to overcome resistance to EGFR tyrosine kinase inhibitors in lung cancer. Cancer Lett 2017; 405: 100-10.
[http://dx.doi.org/10.1016/j.canlet.2017.07.023] [PMID: 28774798]

[38]   Yang JC, Shih JY, Su WC, *et al.* Afatinib for patients with lung adenocarcinoma and epidermal growth factor receptor mutations (LUX-Lung 2): a phase 2 trial. Lancet Oncol 2012; 13(5): 539-48.
[http://dx.doi.org/10.1016/S1470-2045(12)70086-4] [PMID: 22452895]

[39]   Sequist LV, Yang JC, Yamamoto N, *et al.* Phase III study of afatinib or cisplatin plus pemetrexed in patients with metastatic lung adenocarcinoma with EGFR mutations. J Clin Oncol 2013; 31(27): 3327-34.
[http://dx.doi.org/10.1200/JCO.2012.44.2806] [PMID: 23816960]

[40]   Wu YL, Zhou C, Hu CP, *et al.* Afatinib *versus* cisplatin plus gemcitabine for first-line treatment of Asian patients with advanced non-small-cell lung cancer harbouring EGFR mutations (LUX-Lung 6): an open-label, randomised phase 3 trial. Lancet Oncol 2014; 15(2): 213-22.
[http://dx.doi.org/10.1016/S1470-2045(13)70604-1] [PMID: 24439929]

[41]   Yang JC, Wu YL, Schuler M, *et al.* Afatinib *versus* cisplatin-based chemotherapy for EGFR mutation-positive lung adenocarcinoma (LUX-Lung 3 and LUX-Lung 6): analysis of overall survival data from two randomised, phase 3 trials. Lancet Oncol 2015; 16(2): 141-51.
[http://dx.doi.org/10.1016/S1470-2045(14)71173-8] [PMID: 25589191]

[42]   Park K, Tan EH, O'Byrne K, *et al.* Afatinib versus gefitinib as first-line treatment of patients with EGFR mutation-positive non-small-cell lung cancer (LUX-Lung 7): a phase 2B, open-label, randomised controlled trial. Lancet Oncol 2016; 17(5): 577-89.
[http://dx.doi.org/10.1016/S1470-2045(16)30033-X] [PMID: 27083334]

[43]   Rosell R, Carcereny E, Gervais R, *et al.* Erlotinib versus standard chemotherapy as first-line treatment for European patients with advanced EGFR mutation-positive non-small-cell lung cancer (EURTAC): a multicentre, open-label, randomised phase 3 trial. Lancet Oncol 2012; 13(3): 239-46.
[http://dx.doi.org/10.1016/S1470-2045(11)70393-X] [PMID: 22285168]

[44]   Fenchel K, Dale SP, Dempke WC. Improved overall survival following tyrosine kinase inhibitor (TKI) treatment in NSCLC-are we making progress? Transl Lung Cancer Res 2016; 5(4): 373-6.
[http://dx.doi.org/10.21037/tlcr.2016.07.01] [PMID: 27652201]

[45]   L. Paz-ares EHT, Zhang L, Hirsh V, *et al.* Afatinib (A) *vs* gefitinib (G) in patients with EGFR mutation-positive (Egfrm+) non-small-cell lung cancer (NSCLC): Overall survival (OS) data from the Phase IIb trial LUX-Lung 7 (LL7) [Abstract No. LBA43]. Annals Of Oncology Presented In European Society For Medical Oncology 2016 Congress, 27

[46]   Hirsh V, Cadranel J, Cong XJ, *et al.* Symptom and quality of life benefit of afatinib in advanced non-small-cell lung cancer patients previously treated with erlotinib or gefitinib: results of a randomized phase IIb/III trial (LUX-Lung 1). J Thorac Oncol 2013; 8(2): 229-37.
[http://dx.doi.org/10.1097/JTO.0b013e3182773fce] [PMID: 23328549]

[47]   Jänne PA, Boss DS, Camidge DR, *et al.* Phase I dose-escalation study of the pan-HER inhibitor, PF299804, in patients with advanced malignant solid tumors. Clin Cancer Res 2011; 17(5): 1131-9.
[http://dx.doi.org/10.1158/1078-0432.CCR-10-1220] [PMID: 21220471]

[48]   Park K, Cho BC, Kim DW, *et al.* Safety and efficacy of dacomitinib in korean patients with KRAS wild-type advanced non-small-cell lung cancer refractory to chemotherapy and erlotinib or gefitinib: a phase I/II trial. J Thorac Oncol 2014; 9(10): 1523-31.
[http://dx.doi.org/10.1097/JTO.0000000000000275] [PMID: 25521398]

[49]   Takahashi T, Boku N, Murakami H, *et al.* Phase I and pharmacokinetic study of dacomitinib (PF-00299804), an oral irreversible, small molecule inhibitor of human epidermal growth factor receptor-1, -2, and -4 tyrosine kinases, in Japanese patients with advanced solid tumors. Invest New Drugs 2012; 30(6): 2352-63.
[http://dx.doi.org/10.1007/s10637-011-9789-z] [PMID: 22249430]

[50]   Jänne PA, Ou SH, Kim DW, *et al.* Dacomitinib as first-line treatment in patients with clinically or molecularly selected advanced non-small-cell lung cancer: a multicentre, open-label, phase 2 trial. Lancet Oncol 2014; 15(13): 1433-41.
[http://dx.doi.org/10.1016/S1470-2045(14)70461-9] [PMID: 25456362]

[51]   Wu YL, Cheng Y, Zhou X, *et al.* Dacomitinib versus gefitinib as first-line treatment for patients with EGFR-mutation-positive non-small-cell lung cancer (ARCHER 1050): a randomised, open-label, phase 3 trial. Lancet Oncol 2017; 18(11): 1454-66.
[http://dx.doi.org/10.1016/S1470-2045(17)30608-3] [PMID: 28958502]

[52]   Ramalingam SS, Blackhall F, Krzakowski M, *et al.* Randomized phase II study of dacomitinib (PF-00299804), an irreversible pan-human epidermal growth factor receptor inhibitor, *versus* erlotinib in patients with advanced non-small-cell lung cancer. J Clin Oncol 2012; 30(27): 3337-44.
[http://dx.doi.org/10.1200/JCO.2011.40.9433] [PMID: 22753918]

[53]   Ramalingam SS, Jänne PA, Mok T, *et al.* Dacomitinib *versus* erlotinib in patients with advanced-stage, previously treated non-small-cell lung cancer (ARCHER 1009): a randomised, double-blind, phase 3 trial. Lancet Oncol 2014; 15(12): 1369-78.
[http://dx.doi.org/10.1016/S1470-2045(14)70452-8] [PMID: 25439691]

[54]   Reckamp KL, Giaccone G, Camidge DR, *et al.* A phase 2 trial of dacomitinib (PF-00299804), an oral, irreversible pan-HER (human epidermal growth factor receptor) inhibitor, in patients with advanced non-small cell lung cancer after failure of prior chemotherapy and erlotinib. Cancer 2014; 120(8): 1145-54.
[http://dx.doi.org/10.1002/cncr.28561] [PMID: 24501009]

[55]   Ellis PM, Shepherd FA, Millward M, *et al.* Dacomitinib compared with placebo in pretreated patients with advanced or metastatic non-small-cell lung cancer (NCIC CTG BR.26): a double-blind, randomised, phase 3 trial. Lancet Oncol 2014; 15(12): 1379-88.
[http://dx.doi.org/10.1016/S1470-2045(14)70472-3] [PMID: 25439692]

[56]   Jänne PA, von Pawel J, Cohen RB, *et al.* Multicenter, randomized, phase II trial of CI-1033, an irreversible pan-ERBB inhibitor, for previously treated advanced non small-cell lung cancer. J Clin Oncol 2007; 25(25): 3936-44.
[http://dx.doi.org/10.1200/JCO.2007.11.1336] [PMID: 17761977]

[57]   Shepherd FA, Rodrigues Pereira J, Ciuleanu T, *et al.* Erlotinib in previously treated non-small-cell lung cancer. N Engl J Med 2005; 353(2): 123-32.
[http://dx.doi.org/10.1056/NEJMoa050753] [PMID: 16014882]

[58]   Wong KK, Fracasso PM, Bukowski RM, *et al.* A phase I study with neratinib (HKI-272), an irreversible pan ErbB receptor tyrosine kinase inhibitor, in patients with solid tumors. Clin Cancer Res 2009; 15(7): 2552-8.
[http://dx.doi.org/10.1158/1078-0432.CCR-08-1978] [PMID: 19318484]

[59]   Sequist LV, Besse B, Lynch TJ, *et al.* Neratinib, an irreversible pan-ErbB receptor tyrosine kinase

inhibitor: results of a phase II trial in patients with advanced non-small-cell lung cancer. J Clin Oncol 2010; 28(18): 3076-83.
[http://dx.doi.org/10.1200/JCO.2009.27.9414] [PMID: 20479403]

[60]   Godin-Heymann N, Ulkus L, Brannigan BW, *et al.* The T790M "gatekeeper" mutation in EGFR mediates resistance to low concentrations of an irreversible EGFR inhibitor. Mol Cancer Ther 2008; 7(4): 874-9.
[http://dx.doi.org/10.1158/1535-7163.MCT-07-2387] [PMID: 18413800]

[61]   Klempner SJ, Bazhenova LA, Braiteh FS, *et al.* Emergence of RET rearrangement co-existing with activated EGFR mutation in EGFR-mutated NSCLC patients who had progressed on first- or second-generation EGFR TKI. Lung Cancer 2015; 89(3): 357-9.
[http://dx.doi.org/10.1016/j.lungcan.2015.06.021] [PMID: 26187428]

[62]   Wu SG, Liu YN, Tsai MF, *et al.* The mechanism of acquired resistance to irreversible EGFR tyrosine kinase inhibitor-afatinib in lung adenocarcinoma patients. Oncotarget 2016; 7(11): 12404-13.
[http://dx.doi.org/10.18632/oncotarget.7189] [PMID: 26862733]

[63]   Campo M, Gerber D, Gainor JF, *et al.* Acquired resistance to first-line afatinib and the challenges of prearranged progression biopsies. J Thorac Oncol 2016; 11(11): 2022-6.
[http://dx.doi.org/10.1016/j.jtho.2016.06.032] [PMID: 27553514]

[64]   Jänne PA, Yang JC, Kim DW, *et al.* AZD9291 in EGFR inhibitor-resistant non-small-cell lung cancer. N Engl J Med 2015; 372(18): 1689-99.
[http://dx.doi.org/10.1056/NEJMoa1411817] [PMID: 25923549]

[65]   Li D, Ambrogio L, Shimamura T, *et al.* BIBW2992, an irreversible EGFR/HER2 inhibitor highly effective in preclinical lung cancer models. Oncogene 2008; 27(34): 4702-11.
[http://dx.doi.org/10.1038/onc.2008.109] [PMID: 18408761]

[66]   Miller VA, Hirsh V, Cadranel J, *et al.* Afatinib versus placebo for patients with advanced, metastatic non-small-cell lung cancer after failure of erlotinib, gefitinib, or both, and one or two lines of chemotherapy (LUX-Lung 1): a phase 2b/3 randomised trial. Lancet Oncol 2012; 13(5): 528-38.
[http://dx.doi.org/10.1016/S1470-2045(12)70087-6] [PMID: 22452896]

[67]   Katakami N, Atagi S, Goto K, *et al.* LUX-Lung 4: a phase II trial of afatinib in patients with advanced non-small-cell lung cancer who progressed during prior treatment with erlotinib, gefitinib, or both. J Clin Oncol 2013; 31(27): 3335-41.
[http://dx.doi.org/10.1200/JCO.2012.45.0981] [PMID: 23816963]

[68]   Godin-Heymann N, Ulkus L, Brannigan BW, *et al.* The T790M "gatekeeper" mutation in EGFR mediates resistance to low concentrations of an irreversible EGFR inhibitor. Mol Cancer Ther 2008; 7(4): 874-9.
[http://dx.doi.org/10.1158/1535-7163.MCT-07-2387] [PMID: 18413800]

[69]   Sos ML, Rode HB, Heynck S, *et al.* Chemogenomic profiling provides insights into the limited activity of irreversible EGFR Inhibitors in tumor cells expressing the T790M EGFR resistance mutation. Cancer Res 2010; 70(3): 868-74.
[http://dx.doi.org/10.1158/0008-5472.CAN-09-3106] [PMID: 20103621]

[70]   Kobayashi Y, Azuma K, Nagai H, *et al.* Characterization of EGFR T790M, L792F, and C797S mutations as mechanisms of acquired resistance to afatinib in lung cancer. Mol Cancer Ther 2017; 16(2): 357-64.
[http://dx.doi.org/10.1158/1535-7163.MCT-16-0407] [PMID: 27913578]

[71]   Yamaoka T, Ohmori T, Ohba M, *et al.* Distinct afatinib resistance mechanisms idenfitied in lung adenocarcinoma harboring an EGFR mutation. Mol Cancer Res 2017; 15(7): 915-28.
[http://dx.doi.org/10.1158/1541-7786.MCR-16-0482] [PMID: 28289161]

[72]   Yang J, Ramalingam SS, Janne PA, Cantarini M, Mitsudomi T. Osimertinib (AZD9291) in pre-treated pts with T790M-positive advanced NSCLC: updated Phase 1 (P1) and pooled Phase 2 (P2) results. J Thorac Oncol 2016; 11: S152-3.

[http://dx.doi.org/10.1016/S1556-0864(16)30325-2]

[73]   Mok T, Ahn MJ, Han JY, *et al.* CNS response to osimertinib in patients (pts) with T790M-positive advanced NSCLC: Data from a randomized phase III trial (AURA3) J Clin Oncol 2017; 35 (15_suppl): 9005-5.

[74]   Soria JC, Ohe Y, Vansteenkiste J, *et al.* Osimertinib in untreated EGFR-mutated advanced non-smal-
-cell lung cancer. N Engl J Med 2018; 378(2): 113-25.
[http://dx.doi.org/10.1056/NEJMoa1713137] [PMID: 29151359]

[75]   Yang JC, Cho BC, Kim DW, *et al.* Osimertinib for patients (pts) with leptomeningeal metastases (LM) from EGFR-mutant non-small cell lung cancer (NSCLC): updated results from the BLOOM study

[76]   Goss G, Tsai CM, Shepherd FA, *et al.* CNS response to osimertinib in patients with T790M-positive advanced NSCLC: pooled data from two phase II trials. Ann Oncol 2018; 29(3): 687-93.
[http://dx.doi.org/10.1093/annonc/mdx820] [PMID: 29293889]

[77]   Lee CK, Novello S, Ryden A, *et al.* Patient-reported symptoms and impact of treatment with osimertinib *vs* chemotherapy for advanced non-small cell lung cancer Ann Oncol 2017; 28 (suppl_2): ii28-51.

[78]   Sequist LV, Soria JC, Goldman JW, *et al.* Rociletinib in EGFR-mutated non-small-cell lung cancer. N Engl J Med 2015; 372(18): 1700-9.
[http://dx.doi.org/10.1056/NEJMoa1413654] [PMID: 25923550]

[79]   Simmons AD, Tsai SJ, Haringsma HJ, *et al.* Insulin-like growth factor 1 (IGF1R)/insulin receptor (INSR) inhibitory activity of rociletinib (CO-1686) and its metabolites in nonclinical models. [abstract]. Proceedings of the 106th Annual Meeting of the American Association for Cancer Research. 2015 Apr 18-22; Philadelphia, PA. 2015.
[http://dx.doi.org/10.1158/1538-7445.AM2015-793]

[80]   Morgillo F, Woo JK, Kim ES, Hong WK, Lee HY. Heterodimerization of insulin-like growth factor receptor/epidermal growth factor receptor and induction of survivin expression counteract the antitumor action of erlotinib. Cancer Res 2006; 66(20): 10100-11.
[http://dx.doi.org/10.1158/0008-5472.CAN-06-1684] [PMID: 17047074]

[81]   Yasushi Goto HN, Mrakami H, Shimizu T, *et al.* ASP8273, A mutant-selective irreversible EGFR inhibitor in patients with NSCLC harboring EGFR activating mutations: preliminary results of first-i-
-human phase I study In Japan J Clin Oncol 2015; 33 (Suppl) Abstr 8014

[82]   Ortiz-Cuaran S, Scheffler M, Plenker D, *et al.* Heterogeneous mechanisms of primary and acquired resistance to third-generation EGFR inhibitors. Clin Cancer Res 2016; 22(19): 4837-47.
[http://dx.doi.org/10.1158/1078-0432.CCR-15-1915] [PMID: 27252416]

[83]   Chabon JJ, Simmons AD, Lovejoy AF, *et al.* Circulating tumour DNA profiling reveals heterogeneity of EGFR inhibitor resistance mechanisms in lung cancer patients. Nat Commun 2016; 7: 11815.
[http://dx.doi.org/10.1038/ncomms11815] [PMID: 27283993]

[84]   Minari R, Bordi P, Tiseo M. Third-generation epidermal growth factor receptor-tyrosine kinase inhibitors in T790M-positive non-small cell lung cancer: review on emerged mechanisms of resistance. Transl Lung Cancer Res 2016; 5(6): 695-708.
[http://dx.doi.org/10.21037/tlcr.2016.12.02] [PMID: 28149764]

[85]   Kim TM, Song A, Kim DW, *et al.* Mechanisms of acquired resistance to AZD9291: a mutation-selective, irreversible EGFR inhibitor. J Thorac Oncol 2015; 10(12): 1736-44.
[http://dx.doi.org/10.1097/JTO.0000000000000688] [PMID: 26473643]

[86]   Ham JS, Kim S, Kim HK, *et al.* Two cases of small cell lung cancer transformation from EGFR mutant adenocarcinoma during AZD9291 treatment. J Thorac Oncol 2016; 11(1): e1-4.
[http://dx.doi.org/10.1016/j.jtho.2015.09.013] [PMID: 26762749]

[87]   Walter AO, Sjin RT, Haringsma HJ, *et al.* Discovery of a mutant-selective covalent inhibitor of EGFR that overcomes T790M-mediated resistance in NSCLC. Cancer Discov 2013; 3(12): 1404-15.

[http://dx.doi.org/10.1158/2159-8290.CD-13-0314] [PMID: 24065731]

[88]   Togashi Y, Masago K, Masuda S, *et al.* Cerebrospinal fluid concentration of gefitinib and erlotinib in patients with non-small cell lung cancer. Cancer Chemother Pharmacol 2012; 70(3): 399-405.
[http://dx.doi.org/10.1007/s00280-012-1929-4] [PMID: 22806307]

[89]   Deng Y, Feng W, Wu J, *et al.* The concentration of erlotinib in the cerebrospinal fluid of patients with brain metastasis from non-small-cell lung cancer. Mol Clin Oncol 2014; 2(1): 116-20.
[http://dx.doi.org/10.3892/mco.2013.190] [PMID: 24649318]

[90]   Thress KS, Paweletz CP, Felip E, *et al.* Acquired EGFR C797S mutation mediates resistance to AZD9291 in non-small cell lung cancer harboring EGFR T790M. Nat Med 2015; 21(6): 560-2.
[http://dx.doi.org/10.1038/nm.3854] [PMID: 25939061]

[91]   cobas EGFR Mutation Test v2. 2016. Available from: http:///www. fda. gov/ Drugs/informationOnDrugs/ApprovedDrugs/ucm504540.htm

[92]   Takeda M, Nakagawa K. Toxicity profile of epidermal growth factor receptor tyrosine kinase inhibitors in patients with epidermal growth factor receptor gene mutation-positive lung cancer. Mol Clin Oncol 2017; 6(1): 3-6.
[http://dx.doi.org/10.3892/mco.2016.1099] [PMID: 28123721]

[93]   Satoh H, Inoue A, Kobayashi K, *et al.* Low-dose gefitinib treatment for patients with advanced non-small cell lung cancer harboring sensitive epidermal growth factor receptor mutations. J Thorac Oncol 2011; 6(8): 1413-7.
[http://dx.doi.org/10.1097/JTO.0b013e31821d43a8] [PMID: 21681118]

[94]   Binder D, Buckendahl AC, Hübner RH, *et al.* Erlotinib in patients with advanced non-small-cell lung cancer: impact of dose reductions and a novel surrogate marker. Med Oncol 2012; 29(1): 193-8.
[http://dx.doi.org/10.1007/s12032-010-9767-x] [PMID: 21327738]

[95]   Chen X, Zhu Q, Liu Y, *et al.* Icotinib is an active treatment of non-small-cell lung cancer: a retrospective study. PLoS One 2014; 9(5): e95897.
[http://dx.doi.org/10.1371/journal.pone.0095897] [PMID: 24836053]

[96]   Lacouture ME, Schadendorf D, Chu CY, *et al.* Dermatologic adverse events associated with afatinib: an oral ErbB family blocker. Expert Rev Anticancer Ther 2013; 13(6): 721-8.
[http://dx.doi.org/10.1586/era.13.30] [PMID: 23506519]

[97]   Rixe O, Franco SX, Yardley DA, *et al.* A randomized, phase II, dose-finding study of the pan-ErbB receptor tyrosine-kinase inhibitor CI-1033 in patients with pretreated metastatic breast cancer. Cancer Chemother Pharmacol 2009; 64(6): 1139-48.
[http://dx.doi.org/10.1007/s00280-009-0975-z] [PMID: 19294387]

[98]   Liao BC, Lin CC, Lee JH, Yang JC. Update on recent preclinical and clinical studies of T790M mutant-specific irreversible epidermal growth factor receptor tyrosine kinase inhibitors. J Biomed Sci 2016; 23(1): 86.
[http://dx.doi.org/10.1186/s12929-016-0305-9] [PMID: 27912760]

[99]   Kim ES, Hirsh V, Mok T, *et al.* Gefitinib *versus* docetaxel in previously treated non-small-cell lung cancer (INTEREST): a randomised phase III trial. Lancet 2008; 372(9652): 1809-18.
[http://dx.doi.org/10.1016/S0140-6736(08)61758-4] [PMID: 19027483]

[100]   Maruyama R, Nishiwaki Y, Tamura T, *et al.* Phase III study, V-15-32, of gefitinib *versus* docetaxel in previously treated Japanese patients with non-small-cell lung cancer. J Clin Oncol 2008; 26(26): 4244-52.
[http://dx.doi.org/10.1200/JCO.2007.15.0185] [PMID: 18779611]

[101]   Lee DH, Park K, Kim JH, *et al.* Randomized Phase III trial of gefitinib *versus* docetaxel in non-small cell lung cancer patients who have previously received platinum-based chemotherapy. Clin Cancer Res 2010; 16(4): 1307-14.
[http://dx.doi.org/10.1158/1078-0432.CCR-09-1903] [PMID: 20145166]

[102] Takeda K, Hida T, Sato T, *et al*. Randomized phase III trial of platinum-doublet chemotherapy followed by gefitinib compared with continued platinum-doublet chemotherapy in Japanese patients with advanced non-small-cell lung cancer: results of a west Japan thoracic oncology group trial (WJTOG0203). J Clin Oncol 2010; 28(5): 753-60.
[http://dx.doi.org/10.1200/JCO.2009.23.3445] [PMID: 20038730]

[103] Gaafar RM, Surmont VF, Scagliotti GV, *et al*. A double-blind, randomised, placebo-controlled phase III intergroup study of gefitinib in patients with advanced NSCLC, non-progressing after first line platinum-based chemotherapy (EORTC 08021/ILCP 01/03). Eur J Cancer 2011; 47(15): 2331-40.
[http://dx.doi.org/10.1016/j.ejca.2011.06.045] [PMID: 21802939]

[104] Han JY, Park K, Kim SW, *et al*. First SIGNAL: first-line single-agent iressa *versus* gemcitabine and cisplatin trial in never-smokers with adenocarcinoma of the lung. J Clin Oncol 2012; 30(10): 1122-8.
[http://dx.doi.org/10.1200/JCO.2011.36.8456] [PMID: 22370314]

[105] Inoue A, Kobayashi K, Maemondo M, *et al*. Updated overall survival results from a randomized phase III trial comparing gefitinib with carboplatin-paclitaxel for chemo-naïve non-small cell lung cancer with sensitive EGFR gene mutations (NEJ002). Ann Oncol 2013; 24(1): 54-9.
[http://dx.doi.org/10.1093/annonc/mds214] [PMID: 22967997]

[106] Zhao H, Fan Y, Ma S, *et al*. Final overall survival results from a phase III, randomized, placebo-controlled, parallel-group study of gefitinib *versus* placebo as maintenance therapy in patients with locally advanced or metastatic non-small-cell lung cancer (INFORM; C-TONG 0804). J Thorac Oncol 2015; 10(4): 655-64.
[http://dx.doi.org/10.1097/JTO.0000000000000445] [PMID: 25546556]

[107] Urata Y, Katakami N, Morita S, *et al*. Randomized phase III study comparing gefitinib with erlotinib in patients with previously treated advanced lung adenocarcinoma: WJOG 5108L. J Clin Oncol 2016; 34(27): 3248-57.
[http://dx.doi.org/10.1200/JCO.2015.63.4154] [PMID: 27022112]

[108] Patil VM, Noronha V, Joshi A, *et al*. Phase III study of gefitinib or pemetrexed with carboplatin in EGFR-mutated advanced lung adenocarcinoma. ESMO Open 2017; 2(1): e000168.
[http://dx.doi.org/10.1136/esmoopen-2017-000168] [PMID: 28761735]

[109] Wu YL, Zhou C, Liam CK, *et al*. First-line erlotinib *versus* gemcitabine/cisplatin in patients with advanced EGFR mutation-positive non-small-cell lung cancer: analyses from the phase III, randomized, open-label, ENSURE study. Ann Oncol 2015; 26(9): 1883-9.
[http://dx.doi.org/10.1093/annonc/mdv270] [PMID: 26105600]

[110] Kawaguchi T, Ando M, Asami K, *et al*. Randomized phase III trial of erlotinib *versus* docetaxel as second- or third-line therapy in patients with advanced non-small-cell lung cancer: Docetaxel and Erlotinib Lung Cancer Trial (DELTA). J Clin Oncol 2014; 32(18): 1902-8.
[http://dx.doi.org/10.1200/JCO.2013.52.4694] [PMID: 24841974]

SUBJECT INDEX

A

B

C

www.ingramcontent.com/pod-product-compliance
Lightning Source LLC
Chambersburg PA
CBHW041704210326
41598CB00007B/528